FOR
DUMMIES
A Wiley Brand
2nd Edition

OFFICIAL
DISCARD

LIBRARY
SONOMA COUNTY

by H. Ward Silver

Ham Radio For Dummies®, 2nd Edition

Published by:
John Wiley & Sons, Inc.
111 River Street
Hoboken, NJ 07030-5774
www.wiley.com

Published simultaneously in Canada

For general information on our other products and services, please contact our Customer Care Department within the U.S. at 877-762-2974, outside the U.S. at 317-572-3993, or fax 317-572-4002. For technical support, please visit www.wiley.com/techsupport.

Wiley publishes in a variety of print and electronic formats and by print-on-demand. Some material included with standard print versions of this book may not be included in e-books or in print-on-demand. If this book refers to media such as a CD or DVD that is not included in the version you purchased, you may download this material at http://booksupport.wiley.com. For more information about Wiley products, visit www.wiley.com.

Library of Congress Control Number: 2013938107

ISBN 978-1-118-59211-3 (pbk); ISBN 978-1-118-59220-5 (ebk); ISBN 978-1-118-59204-5 (ebk)

Manufactured in the United States of America

10 9 8 7 6 5 4 3 2 1

Contents at a Glance

Table of Contents

Part IV: Building and Operating a Station That Works... 215

Introduction

You may have come across ham radio in any number of ways. Maybe you browsed a website that has some ham radio content, such as a project on a do-it-yourself website or in a YouTube video. Hams also play key roles in movies such as *Contact* and on television series such as *Last Man Standing,* in which Tim Allen plays a ham who has stations in his basement and at work. Several cartoon characters on *The Simpsons* are hams, too! You'll discover hams and ham radio by reading about them in newspaper and magazine articles, by seeing them in action performing emergency communications services, by interacting with a teacher or professor, or maybe by talking to a friend or relative who enjoys the hobby. Interestingly enough, ham radio has room for all these activities. Ham radio includes a mad scientist or two, but most hams are just like you.

The traditional image of ham radio is of a room full of vacuum tubes, gears, flickering needles, and Morse code equipment, but today's hams have many more options to try. Although the traditional shortwave bands are certainly crowded with ham signals hopping around the planet, hams now transmit data and pictures through the airwaves; use the Internet, lasers, and microwave transmitters; and travel to unusual places high and low to make contact, even to and from the International Space Station.

Simply stated, ham radio provides the broadest and most powerful wireless communications capability available to any private citizen anywhere in the world. Because the world's citizens are craving ever-closer contact and hands-on experiences with technology of all sorts, ham radio is attracting attention from people like you. The hobby has never had more to offer and shows no sign of slowing its expansion into new wireless technologies. (Did I say wireless? Think extreme wireless!)

About This Book

I wrote *Ham Radio For Dummies,* 2nd Edition, for beginning hams. If you've just become interested in ham radio, you'll find plenty of information here on what the hobby is all about and how to go about joining the fun by discovering the basics and getting a license. Many books about ham radio's technical and operating specialties are available, but this book introduces them briefly so you can get up to speed as quickly as possible.

If you've already received your license, congratulations! This book helps you change from a listener to a doer. Any new hobby, particularly a technical one, can be intimidating to newcomers. By keeping *Ham Radio For Dummies* handy in your radio shack, you'll be able to quickly understand what you hear on the airwaves. I cover the basics of getting a station put together properly and the fundamentals of on-the-air behavior. Use this book as your personal radio buddy, and soon, you'll be making contacts with confidence.

You can read this book in any order. Feel free to browse and flip through the pages to any section that catches your interest. The sidebars and icons are there to support the main story of each chapter, but you can skip them and come back to them later.

The book has five parts. Parts I and II are for readers who are getting interested in ham radio and preparing to get a license. Parts III and IV explain how to set up a station, get on the air, and make contact with other hams. Part V is the Part of Tens (familiar to all *For Dummies* readers), which presents a few tips and secrets of ham radio. The appendixes consist of an extensive glossary and a long list of excellent references — both online and off — for you to use as you branch out and expand your ham radio career.

Within this book, you may note that some web addresses break across two lines of text. If you're reading this book in print and want to visit one of these web pages, simply key in the address exactly as it's noted in the text, pretending that the line break doesn't exist. If you're reading this book as an e-book, you've got it easy; just click the web address to be taken directly to the web page.

Foolish Assumptions

In writing this book, I made some assumptions about you. You don't have to know a single thing about ham radio or its technology to enjoy *Ham Radio For Dummies,* 2nd Edition, and you definitely don't need to be an electrical engineer to enjoy this book.

But I ask two things of you:

- You have an interest in ham radio.
- You know how to use a computer well enough to surf the web.

Due to the broad nature of ham radio, I couldn't include everything in this book. (Also, if I'd done that, you wouldn't be able to lift it.) But I steer you in the direction of additional resources, including websites, that will help you follow this book with current information and explanations.

Icons Used in This Book

While you're reading, you'll notice icons that point out special information. Here are the icons I use and what they mean.

This icon points out easier, shorter, or more direct ways of doing something.

This icon goes with information that helps you operate effectively and avoid technical bumps in the road.

This icon signals when I show my techie side. If you don't want to know the technical details, skip paragraphs marked with this icon.

Whenever I could think of a common problem or "oops," you see this icon. Before you become experienced, it's easy to get hung up on little things.

This icon lets you know that some regulatory, safety, or performance issues are associated with the topic of discussion. Watch for this icon to avoid common gotchas.

Beyond the Book

Appendix B provides a long list of great resources for you, so definitely start there when you have questions about topics that aren't covered in the rest of the book.

Next, you'll want to keep your eCheat Sheet handy; that's where I've collected the information you're most likely to need on a day-to-day basis while you're building up your ham radio know-how. You'll find a summary of your Technician (and soon-to-be General) class license privileges and other stuff you'll want in your shack or at your fingertips. The eCheat Sheet is available at www.dummies.com/cheatsheet/hamradio.

Finally, if you browse to www.dummies.com/extras/hamradio, you'll find additional online material: useful stuff that you'll be glad you can access online from anywhere.

Where to Go from Here

If you're not yet a ham, I highly recommend that you find your most comfortable chair and read Parts I and II, where you can discover the basics about ham radio and solidify your interest. If you're a licensed ham, browse through Parts III and IV to find the topics that interest you most. Take a look at the appendixes to find out what information is secreted away back there for when you need it in a hurry.

For all my readers, welcome to *Ham Radio For Dummies,* 2nd Edition. I hope to meet you on the air someday!

Part I

Getting Started with Ham Radio

getting started
with
ham
radio

Visit `www.dummies.com` for more great *For Dummies* content online.

In this part . . .

- Get acquainted with ham radio — what it is and how hams contact one another.
- Find out about the basic technologies that form the foundation of ham radio.
- Discover how hams interact with the natural world to communicate across town and around the world.
- Get acquainted with the various types of ham communities: on the air, online, and in person.
- Visit `www.dummies.com` for more great *For Dummies* content online.

Chapter 1

Getting Acquainted with Ham Radio

In This Chapter

- Becoming part of ham radio
- Exploring ham radio
- Making contact via ham radio
- Constructing your own radio shack

Ham radio invokes a wide range of visions. Maybe you have a mental image of a ham radio operator (or *ham)* from a movie or newspaper article. But hams are a varied lot — from go-getter emergency communicators to casual chatters to workshop tinkerers. Everyone has a place, and you do too.

Hams employ all sorts of radios and antennas using a wide variety of signals to communicate with other hams across town and around the world. They use ham radio for personal enjoyment, for keeping in touch with friends and family, for emergency communications and public service, and for experimenting with radios and radio equipment. They communicate by using microphones, telegraph or Morse keys, computers, cameras, lasers, and even their own satellites.

Hams meet on the air and in person, in clubs and organizations devoted to every conceivable purpose. Hams run special flea markets and host conventions large and small. Some hams are as young as 6 years old; others are centenarians. Some have a technical background; most do not. One thing that all these diverse people share, however, is an interest in radio that can express itself in many ways.

This chapter gives you an overview of the world of ham radio and shows you how to become part of it.

Tuning into Ham Radio Today

Hams enjoy three aspects of ham radio: technology, operations, and social. Your interest in the hobby may be technical, you may want to use ham radio for a specific purpose, or you may just want to join the fun. All are perfectly valid reasons for getting a ham radio license.

Using electronics and technology

Ham radio lets you work closely with electronics and technology (see Chapter 2). Just transmitting and receiving radio signals can be an electronics-intensive endeavor. By opening the hood on the ham radio hobby, you're gaining experience with everything from basic electronics to cutting-edge wireless techniques. Everything from analog electronics to the latest in digital signal processing and computing technology is available in ham radio. Whatever part of electronic and computing technology you enjoy most, it's all used in ham radio somewhere . . . and sometimes, all at the same time!

In this section, I give you a quick look at what you can do with technology.

You don't have to know everything that there is to know, though. I've been in the hobby for more than 40 years, and I've never met anyone who's an expert on everything.

Design and build

Just as an audiophile might, you can design and build your own equipment or assemble a station from factory-built components. All the components you need to take either path are widely available in stores and on the web. The original do-it-yourself (DIY) technical hobbyists, hams delight in the ethic known as *homebrewing,* helping one another build and maintain stations.

Create hybrid software and systems

You can also write your own software and create brand-new systems that are novel hybrids of radio and the Internet. Hams developed packet radio, for example, by adapting data transmission protocols used in computer networks to amateur radio links. Packet radio is now widely used in many commercial applications.

Also, the combination of GPS radiolocation technology with the web and amateur mobile radios resulted in the Automatic Packet Reporting System (APRS), which is now widely used. For more information about these neat systems, see Parts III and IV of this book.

Digitize your radio

Voice and Morse-code communications are still the most popular technologies that hams use to talk to one another, but computer-based digital operation is gaining fast. The most common home-station configuration today is a combination of computer and radio. The newest radios are based on software-defined radio (SDR) technology, which allows the radio to change itself on the fly to adapt to new conditions or perform new functions, as you see in Part IV.

Experiment with antennas

Besides being students of equipment and computers, hams are students of antennas and *propagation,* which is the means by which radio signals bounce around from place to place. Hams take an interest in solar cycles and sunspots and in how they affect the Earth's ionosphere. For hams, weather takes on new importance, generating static or fronts along which radio signals can sometimes travel long distances. Antennas launch signals to take advantage of all this propagation, so they provide a fertile universe for station builders and experimenters.

Antenna experimentation and computer modeling is a hotbed of activity for hams. New designs are created every day, and hams have contributed many advances and refinements to the antenna designer's art. Antenna systems range from small patches of printed circuit-board material to multiple towers festooned with large rotating arrays. All you need to start growing your own antenna farm are some wire, a feed line, and some basic tools. I give you the full lowdown in Chapter 12.

Enhance other hobbies

Hams use radio technology in support of hobbies such as radio control (R/C), model rocketry, and ballooning. Hams have special frequencies for R/C operation in their 6 meter band, away from the crowded unlicensed R/C frequencies. Miniature ham radio video transmitters (described in Chapter 11) frequently fly in model aircraft, rockets, and balloons, beaming back pictures and location information from altitudes of hundreds or thousands of feet. Ham radio data links are also used in support of astronomy, aviation, auto racing and rallies, and many other pastimes.

The radio in your pocket

You already use a radio to transmit all the time, although you probably don't think of it that way. Your mobile phone is actually a very sophisticated, low-power portable radio! You don't have to get a license to use it, of course; the phone company takes care of that. Nevertheless, your phone is really a radio, transmitting and receiving radio waves that are very similar to some of the radio waves that hams use. As you find out more about ham radio, you'll also find out more about radio waves in general, and you'll begin to look at your mobile phone in a whole new light.

Joining the ham radio community

Hams like to meet in person as well as on the radio. This section discusses a few ways to get involved.

Clubs

Membership in at least one radio club is part of nearly every ham's life. In fact, in some countries, you're required to be a member of a club before you can even get a license.

Chapter 3 shows you how to find and join clubs, which are great sources of information and assistance for new hams.

A particularly beneficial relationship exists between ham radio operators and stamp collectors, or *philatelists.* Hams routinely exchange postcards called QSLs (ham shorthand for *received and understood)* with their call signs, information about their stations, and (often) colorful graphics or photos. Stamp-collecting hams combine the exchange of QSLs with collecting by sending the cards around the world with colorful local stamps or special postmarks. Foreign hams return the favor by sending stamps of their own. The arrival of those cheerful red-and-blue airmail envelopes from exotic locations is always a special treat. You can find more information about the practice of QSLing in Chapter 14.

Hamfests and conventions

Two other popular types of gatherings are hamfests and conventions. A *hamfest* is a ham radio flea market where hams bring their electronic treasures for sale or trade. Some hamfests are small get-togethers held in parking lots on Saturday mornings; others attract thousands of hams from all over the world and last for days, even in these days of eBay!

Hams also hold conventions with a variety of themes, ranging from public service to DX (see "DXing, contests, and awards," later in this chapter) to low-power operating. Hams travel all over the world to attend conventions so they can meet friends formerly known only as voices and call signs over the crackling radio waves.

Emergency teams

Because of their numbers and their reliance on minimal infrastructure, hams bounce back quickly when a natural disaster or other emergency makes communications over normal channels impossible. Hams organize themselves into local and regional teams that practice responding to a variety of emergency needs, working to support relief organizations such as the American Red Cross and the Salvation Army, as well as public-safety agencies such as police and fire departments.

Fall is hurricane season in North America, so every year, ham emergency teams gear up for these potentially devastating storms. These teams staff an amateur station at the National Hurricane Center in Florida (`www.fiu.edu/orgs/w4ehw`) and keep The Hurricane Watch Net busy on 14.325 MHz (`www.hwn.org`).

After big storms of all types, hams are the first messengers from the affected areas, with many more operators around the country standing by to relay their messages and information. During and after Hurricane Sandy in October 2012, for example, hams were on the job providing communications at emergency operations centers and in the field. Government agencies had to focus on coordinating recovery and rescue efforts, so the hams trained as emergency response teams helped them by handling health-and-welfare messages, performing damage assessments, and providing point-to-point coverage until normal communications system came back to life in the following days. To find out more about providing emergency communications and public service, see Chapter 10.

On the last full weekend of June, hams across the United States engage in an annual emergency-operations exercise called Field Day, which allows hams to practice operating under emergency conditions. An amateur emergency team or station probably operates in your town or county; go visit! The American Radio Relay League (ARRL), which is a national association for amateur radio, provides a Field Day Station Locator web page (`www.arrl.org/field-day-locator`) that shows you how to find the team or station nearest you.

Community events

Hams provide assistance for more than just emergencies. Wherever you find a parade, festival, marathon, or other opportunity to provide communications services, you may find ham radio operators helping out. In fact, volunteering for community events is great training for emergencies!

Making contacts

If you were to tune a radio across the ham bands, what would you hear hams doing? They're talking to other hams, of course. These chats, called *contacts,* consist of everything from simple conversations to on-the-air meetings to contesting (discussed later in this chapter). This section gives you a broad overview of contacts; for more info, see "Communicating with Ham Radio," later in this chapter. I discuss contacts in depth in Chapter 8.

Ragchews

By far the most common type of activity for hams is just engaging in conversation, which is called *chewing the rag.* Such contacts are called *ragchews.* Ragchews take place across continents or across town. You don't have to know another ham to have a great ragchew with him or her; ham radio is a friendly hobby with little class snobbery or distinctions. Just make contact, and start talking! Find out more about ragchews in Chapter 9.

The origins of the word *ragchew* are fairly clear. The phrase *chewing the rag* was well known even in the late Middle Ages. *Chew* was slang for *talk,* and *rag,* derived from *fat,* was a reference to the tongue. Thus, people began to use *chewing the rag* to describe to conversations, frequently those that took place during meals. Later, telegraph operators picked up that use, and hams picked it up from telegraphers. Because most of ham radio is in fact conversation, ragchewing has been part of radio since its earliest days.

Nets

Nets (an abbreviation for *networks)* are organized on-the-air meetings scheduled for hams who have a shared interest or purpose. Here are some of the types of nets you can find:

- **Traffic:** These nets are part of the North American system that moves text messages, or *traffic,* via ham radio. Operators meet to exchange *(relay)* messages, sometimes handling dozens in a day. Messages range from mundane communications to emergency health-and-welfare transmissions.
- **Emergency service:** Most of the time, these nets meet only for training and practice. When disasters or other emergencies strike, hams organize around these nets to provide crucial communications into and out of the stricken areas until normal links are restored.
- **Technical service:** These nets are like radio call-in programs; stations call in with specific questions or problems. The net control station may help, but more frequently, one of the listening stations contributes the answer. Many technical-service nets are designed specifically to assist new hams.

- **Swap:** Between the in-person hamfests and flea markets, in many areas a weekly local swap net allows hams to list items for sale or things they need. A net control station moderates the process, putting interested parties in contact with each other; then the parties generally conduct their business over the phone or by e-mail.
- **Mailbox:** If you could listen to Internet systems make contact and exchange data, a mailbox net is what they'd sound like. Instead of transmitting ones and zeroes as voltages on wires, hams use tones. Mailbox nets use computer radio systems that monitor a single frequency all the time so that others can connect to it and send or retrieve messages. Mailboxes are used for emergency communications and for travel where the Internet isn't available.

DXing, contests, and awards

Hams like engaging in challenging activities to build their skills and station capabilities. Following are a few of the most popular activities:

- **DX:** In the world of ham radio, *DX* stands for *distance,* and the allure of making contacts ever more distant from one's home station has always been part of the process. Hams compete to contact faraway stations and to log contacts with every country. They especially enjoy the thrill of contacting exotic locations, such as expeditions to uninhabited islands and remote territories, and making friends in foreign countries. When conditions are right and the band is full of foreign accents, succumbing to the lure of DX is easy.
- **Contests:** Contests are ham radio's version of a contact sport. The point is to make as many contacts as possible during the contest period by sending and receiving as many short messages as possible — sometimes thousands. These exchanges are related to the purpose of the contest: to contact a specific area, use a certain band, find a special station, or just contact the most people.
- **Awards:** Thousands of awards are available for various operating accomplishments, such as contacting different countries or states.
- **Special-event stations:** These temporary stations are on the air for a short time to commemorate or celebrate an event or location, often with a special or collectible call sign. In December 2012, for example, the Marconi Cape Cod Radio Club set up a special temporary station at the location of Marconi's Wellfleet trans-Atlantic operations. Find out more on the club's Facebook page by searching for *KM1CC - Marconi Cape Cod Radio Club.*

If you enjoy the thrill of the chase, go to Chapter 11 to find out more about all these activities.

Hams around the world

Where are the hams in this big world? The International Amateur Radio Union (IARU; www.iaru.org) counts about 160 countries with a national radio society. Counting all the hams in all those countries is difficult, because in some countries, amateur stations and operators have separate licenses. Nevertheless, the best estimate is that there are about 2 million amateur station licenses worldwide and perhaps twice that many amateur operator licenses. The United States alone had more than 700,000 hams as of 2012 — the most ever.

You may not be surprised to hear that China has the fastest-growing amateur population; Thailand and India aren't far behind.

Roaming the World of Ham Radio

Although the United States has a large population of hams, the amateur population in Europe is growing by leaps and bounds, and Japan has an even larger amateur population. With more than 3 million hams worldwide, very few countries are without an amateur (see the nearby sidebar "Hams around the world"). Ham radio is alive and well around the world. Tune the bands on a busy weekend, and you'll see what I mean!

Hams are required to have licenses, no matter where they operate. (I cover all things licensing in Part II of this book.) The international agency that manages radio activity is the International Telecommunication Union (ITU; `www.itu.int/home`). Each member country is required to have its own government agency that controls licensing inside its borders. In the United States, hams are part of the Amateur Radio Service (`http://transition.fcc.gov/pshs/services/amateur.html`), which is regulated and licensed by the Federal Communications Commission (FCC). Outside the United States, amateur radio is governed by similar rules and regulations.

Amateur radio licenses in America are granted by the FCC, but the tests are administered by other hams acting as Volunteer Examiners (VEs). I discuss VEs in detail in Chapter 4. Classes and testing programs are often available through local clubs (refer to "Clubs," earlier in this chapter).

Because radio signals know no boundaries, hams have always been in touch across political borders. Even during the Cold War, U.S. and Soviet hams made regular contact, fostering long personal friendships and international goodwill. Although the Internet makes global communications easy, chatting over the airwaves to someone in another country is exciting and creates a unique personal connection.

Since the adoption of international licensing regulations, hams have operated in many countries with minimal paperwork. A ham from a country that's a party to the international license recognition agreement known as CEPT (an international treaty that enables countries to recognize one another's amateur licenses) can use his or her home license to operate within any other CEPT country. The ARRL provides a lot of useful material about international operating at `www.arrl.org/international-regulatory`.

Communicating with Ham Radio

Though you make contacts for different purposes (such as a chat, an emergency, a net, or a contest), most contacts follow the same structure:

1. You place a call to someone or respond to someone else's call.
2. You and the other operator exchange names, information about where you're located, and the quality of your signal so you can gauge transmitting conditions.
3. If the purpose of the contact is to chat, proceed to chat.

   You might talk about how you constructed your station, what you do for a living, your family, and your job, for example.

Except for the fact that you and the other ham take turns transmitting, and except that this information is converted to radio waves that bounce off the upper atmosphere, making a contact is just like talking to someone you meet at a party or convention. You can hold the conversation by voice, by Morse code, or by keyboard (using a computer as intermediaries to the radios). You won't find a great purpose in the average contact except a desire to meet another ham and see where your radio signal can be heard.

A question that I'm frequently asked about ham radio is "How do you know where to tune for a certain station?" Usually, my answer is "You don't!" Ham radio operators don't have specific frequency assignments or channel numbers. This situation is a good news/bad news situation. The good news is that ham radio gives you unparalleled flexibility to make and maintain communications under continually changing circumstances. The bad news is that making contact with one specific station is hard because you may not know when or on what frequency to call. Hams have found many workarounds for the latter problem, however; the result is an extraordinarily powerful and adaptive communications service.

Building a Ham Radio Shack

For me, the term *radio shack* conjures visions more worthy of a mad scientist's lab than of a modern ham station. Your radio shack, however, is simply the place you keep your radio and ham equipment. The days of bulbous vacuum tubes, jumping meters, and two-handed control knobs are in the distant past.

For some hams, the entire shack consists of a handheld radio or two. Other hams operate on the go in a vehicle. Cars make perfectly good shacks, but most hams have a spot somewhere at home that they claim for a ham radio.

I discuss building and running your own radio shack in Part IV of this book. For now, though, here's what you can find in a ham shack:

- **The rig:** The offspring of the separate receiver and transmitter of yore, the modern radio, or *rig,* combines both devices in a single compact package about the size of a large satellite TV receiver. Like its ancestors, the rig has a large tuning knob that controls the frequency, but state-of-the-art displays and computer screens replace the dials and meters.
- **Computer:** Most hams today have at least one computer in the shack. Computers now control many radio functions, including keeping records of contacts. Digital data communications simply wouldn't be possible without them. Some hams use more than one computer at a time.
- **Mobile/base rig:** For operating through local repeater stations, hams may use a handheld radio, but in their shacks, they use a more-capable radio. These units, which are about the size of hardcover books, can be used as either mobile or base rigs.
- **Microphones, keys, and headphones:** Depending on the shack owner's preferences, you'll see a couple (or more) of these important gadgets, the radio's true user interface. Microphones and keys range from imposing and chrome-plated to miniaturized and hidden. The old Bakelite headphones, or *cans,* are also a distant memory (which is good; they hurt my ears!), replaced by lightweight, comfortable, hi-fi designs.
- **Antennas:** In the shack, you'll find switches and controllers for antennas that live outside the shack. A ham shack tends to sprout antennas ranging from vertical whips the size of pencils to wire antennas stretched through the trees and supersize directional beam antennas held high in the air on steel towers. See Chapter 12 for more info on antennas.
- **Cables and feed lines:** Look behind, around, or under any piece of shack equipment, and you find wires. Lots of them. The radio signals pipe through fat, round black cables called *coaxial,* or *coax.* Power is supplied by colored wires not terribly different in size from house wiring. I cover cables and feed lines in detail in Chapter 12.

The modern ham shack is as far removed from the homebrewed breadboards in the backyard shed as a late-model sedan is from a Model T. You can see examples of several shacks, including mine, in Chapter 13.

Where did the phrase *radio shack* come from? Back in the early days of radio, the equipment was highly experimental and all home-built, requiring a nearby workshop. In addition, the first transmitters used a noisy spark to generate radio waves. The voltages were high, and the equipment was often somewhat a work in progress, so the radio hobbyists often found themselves banished from the house proper. Thus, many early stations were built in a garage or tool shed. The term *shack* carries through today as a description of the state of order and cleanliness in many a ham's lair.

Ham: Not just for sandwiches anymore

Everyone wants to know the meaning of the word *ham,* but as with many slang words, the origin is murky. Theories abound, ranging from the initials of an early radio club's operators to the use of a meat tin as a natural sound amplifier. Of the many possibilities, the following theory seems to be the most believable.

"*Ham:* a poor operator" was used in telegraphy even before radio. The first wireless operators were landline telegraphers who brought with them their language and the traditions of their much older profession. Government stations, ships, coastal stations, and the increasingly numerous amateur operators all competed for signal supremacy in one another's receivers. Many of the amateur stations were very powerful and could effectively jam all the other operators in the area. When this logjam happened, frustrated commercial operators would send the message "THOSE HAMS ARE JAMMING YOU." Amateurs, possibly unfamiliar with the real meaning of the term, picked it up and wore it with pride. As the years advanced, the original meaning completely disappeared.

Chapter 2

Getting a Handle on Ham Radio Technology

In This Chapter

- Discovering radio waves
- Stocking your shack with gadgets
- Understanding the effects of nature on ham radio

Ham radio covers a lot of technological territory — which is one of its most attractive features. To get the most out of ham radio, you need to have a general understanding of the technology that makes ham radio work.

In this chapter, I cover the most common terms and ideas that form the foundation of ham radio. If you want, skip ahead to read about what hams do and how we operate our radios; then come back to this chapter when you need to explore a technical idea.

Exploring the Fundamentals of Radio Waves

Understanding ham radio (or any type of radio) is impossible without also having a general understanding of the purpose of radio: to send and receive information by using radio waves.

Radio waves are just another form of light that travels at the same speed; 186,000 miles per second. Radio waves can get to the Moon and back in 2½ seconds or circle the Earth in ⅐ second.

The energy in a radio wave is partly electric and partly magnetic, appearing as an *electric field* and a *magnetic field* wherever the wave travels. (A *field* is just energy stored in space in one form or another, like a gravitational field that you experience as weight.) These fields make charged particles — such as the electrons in a wire — move in sync with the radio wave. These moving electrons are a *current,* just like in an AC power cord except that they form a radio current that your radio receiver turns into, say, audible speech.

This process works in reverse to create radio waves. *Transmitters* cause electrons to move so that they, in turn, create the radio waves. *Antennas* are just structures in which the electrons move to create and launch radio waves into space. The electrons in an antenna also move in response to radio waves from other antennas. In this way, energy is transferred from moving electrons at one station to radio waves and back to moving electrons at the other station.

Frequency and wavelength

The radio wave–electron relationship has a wrinkle: The fields of the radio wave aren't just one strength all the time; they *oscillate* (vary between a positive and a negative value) the way that a vibrating string moves above and below its stationary position. The time that a field's strength takes to go through one complete set of values is called a *cycle.* The number of cycles in one second is the *frequency* of the wave, measured in *hertz* (abbreviated Hz).

Here's one other wrinkle: The wave is also moving at the speed of light, which is constant. If you could watch the wave oscillate as it moved, you'd see that the wave always moves the same distance — one *wavelength* — in one cycle (see Figure 2-1). The higher the wave's frequency, the faster a cycle completes and the less time it has to move during one cycle. High-frequency waves have short wavelengths, and low-frequency waves have long wavelengths.

If you know a radio wave's frequency, you can figure out the wavelength because the speed of light is always the same. Here's how:

Wavelength = Speed of light / Frequency of the wave

Wavelength in meters = 300,000,000 / Frequency in hertz

Similarly, if you know how far the wave moves in one cycle (the wavelength), you also know how fast it oscillates because the speed of light is fixed:

Frequency in hertz = 300,000,000 / Wavelength in meters

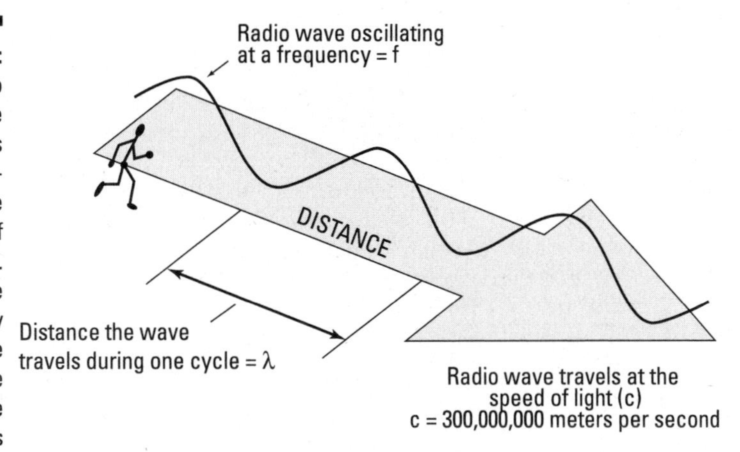

Figure 2-1: As a radio wave travels, its fields oscillate at the frequency of the signal. The distance covered by the wave during one complete cycle is its wavelength.

Courtesy American Radio Relay League

Frequency is abbreviated as *f*, the speed of light as *c*, and wavelength as the Greek letter lambda (λ), leading to the following simple equations:

$$f = c / \lambda \text{ and } \lambda = c / f$$

The higher the frequency, the shorter the wavelength, and vice versa.

Radio waves oscillate at frequencies between the upper end of human hearing at about 20 kilohertz, or kHz (*kilo* is the metric abbreviation meaning 1,000), on up to 1,000 gigahertz, or GHz (*giga* is the metric abbreviation meaning 1 billion). They have corresponding wavelengths from hundreds of meters at the low frequencies to a fraction of a millimeter (mm) at the high frequencies.

The most convenient two units to use in thinking of radio wave frequency (RF) and wavelength are megahertz (MHz; *mega* means 1 million) and meters (m). The equation describing the relationship is much simpler when you use MHz and m:

$$f = 300 / \lambda \text{ in m and } \lambda = 300 / f \text{ in MHz}$$

If you aren't comfortable with memorizing equations, an easy way to convert frequency and wavelength is to memorize just one combination, such as 300 MHz and 1 meter or 10 meters and 30 MHz. Then use factors of ten to move in either direction, making frequency larger and wavelength smaller as you go.

The radio spectrum

The range, or *spectrum,* of radio waves is very broad (see Figure 2-2). Tuning a radio receiver to different frequencies, you hear radio waves carrying all kinds of different information. These radio waves are called *signals.* Signals are grouped by the type of information they carry in different ranges of frequencies, called *bands.* AM broadcast-band stations, for example, transmit signals with frequencies between 550 and 1700 kHz (550,000 and 1,700,000 hertz, or 0.55 and 1.7 MHz). That's what the numbers on a radio dial mean — 550 for 550 kHz and 1000 for 1000 kHz, for example. Bands help you find the type of signals you want without having to hunt over a wide range.

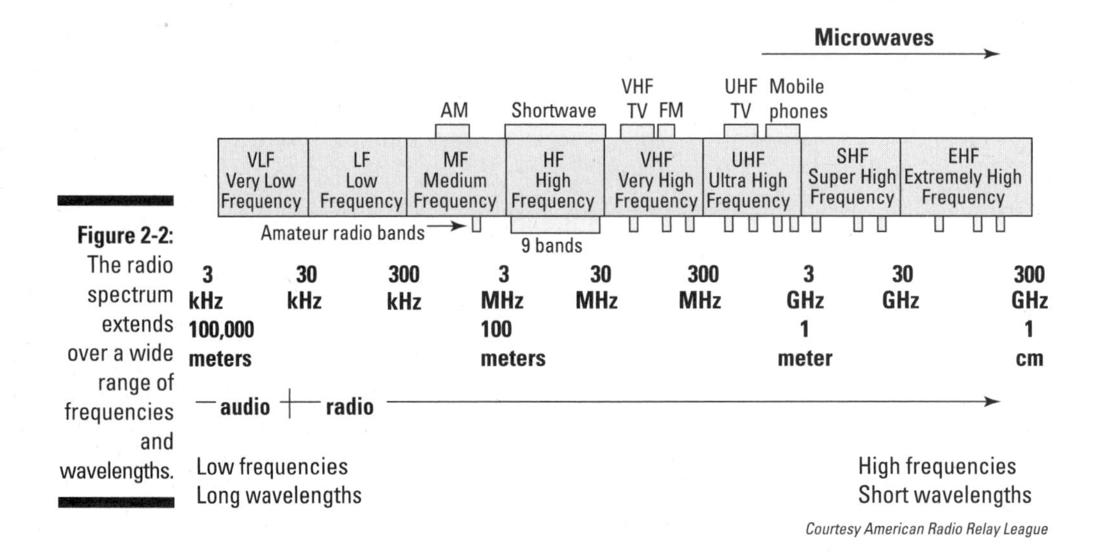

Figure 2-2: The radio spectrum extends over a wide range of frequencies and wavelengths.

The different users of the radio spectrum are called *services,* such as the Broadcasting Service or the Amateur Radio Service. Each service gets a certain amount of spectrum to use, called a *frequency allocation.* Amateur radio, or ham radio, has quite a number of allocations sprinkled throughout the radio spectrum. Hams have access to many small bands in the MF through Microwave regions; I get into the exact locations of the ham radio bands in Chapter 8.

Radio waves at different frequencies act differently in the way they travel, and they require different techniques to transmit and receive. Because waves of similar frequencies tend to have similar properties, the radio spectrum hams use is divided into five segments:

- **Medium Frequency (MF):** Frequencies from 300 kHz to 30 MHz. This segment — the traditional shortwave band — includes AM broadcasting and one ham band. Hams may soon gain access to a pair of bands in this range as the rulemaking process proceeds.
- **Shortwave or High Frequency (HF):** Frequencies from 3 to 30 MHz. This segment — the traditional shortwave band — includes shortwave broadcasting; nine ham radio bands; and ship-to-shore, ship-to-ship, military, and Citizens Band users.
- **Very High Frequency (VHF):** Frequencies from 30 MHz to 300 MHz. This segment includes TV channels 2 through 13, FM broadcasting, three ham bands, public safety and commercial mobile radio, and military and aviation users.
- **Ultra High Frequency (UHF):** Frequencies from 300 MHz to 1 GHz. This segment includes TV channels 14 and higher, two ham bands, cellular phones, public safety and commercial mobile radio, and military and aviation users.
- **Microwave:** Frequencies above 1 GHz. This segment includes GPS; digital wireless telephones; Wi-Fi wireless networking; microwave ovens; eight ham bands; satellite TV; and numerous public, private, and military users.

Because a radio wave has a specific frequency and wavelength, hams use the terms *frequency* and *wavelength* somewhat interchangeably. (The 40 meter and 7 MHz ham bands are the same thing, for example.) I use both terms interchangeably in this book so that you become used to interchanging them as hams are expected to do.

Getting to Know Basic Ham Radio Gadgetry

Although the occasional vacuum-tube radio still glows in an antique-loving ham's station, today's ham radios are sleek, microprocessor-controlled communications centers, as you see in this section.

Basic station

The basic radio is composed of a *receiver* combined with a *transmitter* to make a *transceiver*, which hams call a *rig*. (Mobile and handheld radios are called rigs too.) If the rig doesn't use AC line power directly, a *power supply* provides the DC voltage and current. Figure 2-3 shows the components and gadgets in a typical basic station.

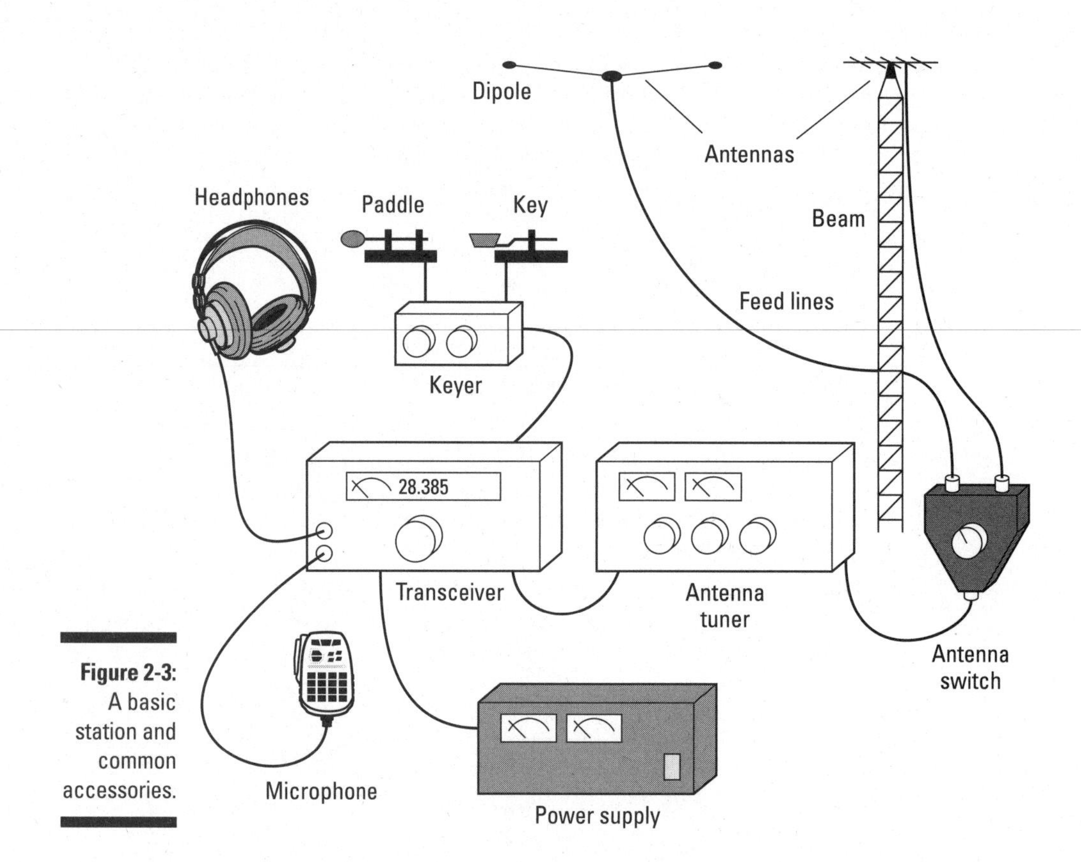

Figure 2-3: A basic station and common accessories.

The radio is connected with a *feed line* to one or more antennas. Two of the most popular antenna types — dipole and beam — are shown in Figure 2-3. A *dipole* is an antenna made from wire and typically connected to its feed line in the middle. A *beam* antenna sends and receives radio waves better in one direction than in others; it's often mounted on a tall pole or tower with a rotator that can point it in different directions. *Antenna switches* allow the operator to select one of several antennas. An *antenna tuner* sits between the antenna/feed line combination and the transmitter, like a vehicle's transmission, to make the transmitter operate at peak efficiency.

You use headphones and a microphone to communicate via the various methods of transmitting speech. If you prefer *Morse code* (also referred to as *CW* for *continuous wave)*, you can use the traditional *straight key* (an old-fashioned Morse code sending device), but more commonly, you use a *paddle* and *keyer,* which are much faster to use than straight keys and require less effort. The paddle looks like a pair of straight keys mounted back to back on their sides. The keyer converts the closing and opening of each paddle lever to strings of dots and dashes electronically.

If you use a computer to generate and receive the individual characters, you're using *digital mode.* When you're using digital mode, you disconnect the microphone and probably the headphones, too, and replace them with connections to the external equipment, as shown in Figure 2-4. A *data interface* passes signals between the radio and computer. For some types of data, a computer can't do the necessary processing, so a *multiprotocol controller* is used. The computer talks to the controller by using a serial RS-232 (COM) or USB port.

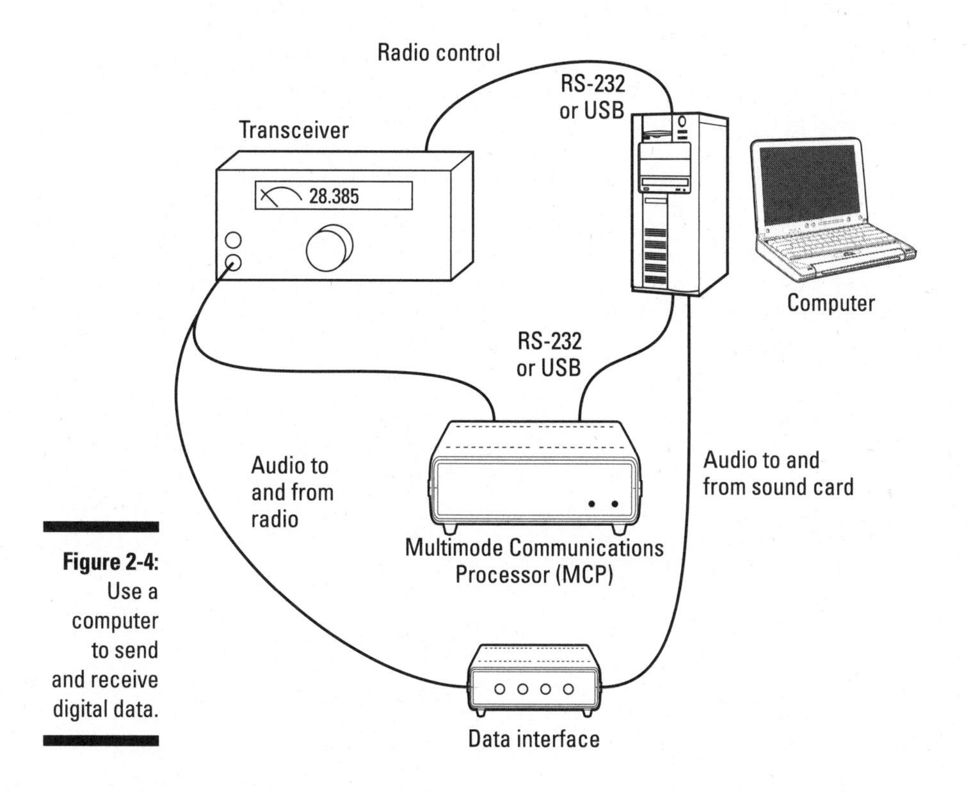

Figure 2-4: Use a computer to send and receive digital data.

Many radios also have an interface (either RS-232 or USB) that allows a computer to control the radio directly. More and more radios are available with Ethernet ports so that they can be connected to a router or home network and operated by remote control.

A lot of programs allow you to control the operating frequency and many other radio settings from a keyboard. Computers can also send and receive Morse code, thereby marrying the hottest 21st-century technology with the oldest form of electronic communications.

Miscellaneous gadgets

Aside from the components that make up your actual operating station, quite a few tools and pieces of equipment make up the rest of your gear.

Feed lines

Two common types of electrical feed lines connect the antennas to the station and carry RF energy between pieces of equipment:

- **Coax:** The most popular type of feed line is coaxial cable, or *coax,* so named because it's constructed of a hollow tube surrounding a central wire. The outer conductor is called the *shield* (or *braid,* if it's made from fine woven wire). The wire in the middle, called the *center conductor,* is surrounded by insulation that holds it right in the center of the cable. The outer conductor is covered by a plastic coating called the *jacket.*
- **Open-wire:** The other kind of feed line is *open-wire* (also called *twin-lead* or *ladder line),* made from two parallel wires. The wires may be exposed, held together with insulating spacers, or covered with plastic insulation.

Feed line measurement

Most radios and antenna tuners have the capability to evaluate the electrical conditions inside the feed line, measured as the *standing wave ratio (SWR).* SWR tells you how well the power from the transmitter is being accepted and turned into radio waves by the antenna. Most radios have a built-in SWR meter. Having a stand-alone SWR meter or SWR bridge is handy when you're working on antennas or operating in a portable situation.

You can also measure feed line conditions by using a *wattmeter,* which measures the actual power flowing back and forth.

SWR meters are inexpensive, but wattmeters are more accurate. These devices typically are used right at the transmitter output.

Filters

Filters are designed to pass or reject ranges of frequencies. Some filters are designed to pass or reject only specific frequencies. Filters can be made from individual or discrete electronic components called *inductors* and *capacitors,* or even from sections of feed line, called *stubs.* Filters come in the following varieties:

- **Feed line:** Use feed line filters to prevent unwanted signals from getting to the radio from the antenna, or vice versa. On transmitted signals, you can use these filters to ensure that unwanted signals from the transmitter aren't radiated, causing interference to others. They also prevent undesired signals from getting to the receiver, where they may compromise receiver performance.

- **Receiving:** Receiving filters are installed inside the radio and usually are made of quartz crystals. These filters remove all but a single desired signal in a receiver. In the latest radio models, filtering is done by software in a special type of digital signal processor (DSP). Filters improve a receiver's *selectivity,* which refers to its capability to receive a single desired signal in the presence of many signals.
- **Audio:** Use audio filters on the receiver output to provide additional filtering capability, rejecting nearby signals or unwanted noise.
- **Notch:** A notch filter removes a narrow range of frequencies, such as a single interfering signal.

Communication technologies

Aside from the equipment, ham radio technology extends to making contacts and exchanging information. You use the following technologies when you use ham radio:

- **Modulation/demodulation:** *Modulation* is the process of adding information to a radio signal so that the information can be transmitted over the air. *Demodulation* is the process of recovering information from a received signal. Ham radios primarily use two kinds of modulation: amplitude modulation (AM) and frequency modulation (FM), almost the same as what you receive on your car radio or home stereo.
- **Modes:** A *mode* is a specific combination of modulation and information. You can choose among several modes when transmitting, including voice, data, video, and Morse code.
- **Repeaters:** *Repeaters* are relay stations that listen on one frequency and retransmit what they hear on a different frequency. Because repeaters are often located on tall buildings, towers, or hilltops, they enable hams to use low-power radios to converse over a wide region. They can be linked by radio or the Internet to extend communication around the world. Repeaters can listen and transmit at the same time — a feature called *duplex* operation.
- **Satellites:** Yes, just like the military and commercial services, hams construct and use their own satellites. (We piggyback on commercial satellite launches; we don't build our own rockets!) Some amateur satellites act like repeaters in the sky; others are used as orbiting digital bulletin boards and e-mail servers.

- **Computer software:** Computers have become big parts of ham radio. Initially, their use was limited to keeping records and making calculations in the shack. Today, they also act as part of the radio, generating and decoding digital data signals, sending Morse code, and controlling the radio's functions. Some radios are almost completely implemented by software running on a PC. Hams have constructed radio-linked computer networks and a worldwide system of e-mail servers accessed by radio.

Hams have always been interested in pushing the envelope when it comes to applying and developing radio technology — one of the fundamental reasons why ham radio exists as a licensed service. Today, ham inventions include such things as creating novel hybrids of radio and other technologies, such as the Internet or GPS radio location. High Speed Multimedia (HSMM), for example (also known as the *hinternet)*, consists of different ham groups working to adapt wireless local area network technology to ham radio. Ham radio is also a hotbed of innovation in antenna design and construction — in short, techie heaven!

Dealing with Mother Nature

Ham radio offers a whole new way of interacting with the natural world around us. Radio waves are affected by the Sun, the characteristics of the atmosphere, and even the properties of ground and water. We may not be able to see these effects with our usual senses, but by using ham radio, we can detect, study, and use them.

Seeing how nature affects radio waves

On their way from Point A to Point B, radio waves have to journey around the Earth and through its atmosphere, encountering a variety of effects:

- **Ground wave propagation:** For local contacts, the radio wave journey along the surface of the Earth is called *ground wave propagation*. Ground wave propagation can support communication up to 100 miles but varies greatly with the frequency being used.
- **Sky wave propagation:** For longer-range contacts, the radio waves must travel through the atmosphere. At HF and sometimes at VHF (refer to "The radio spectrum," earlier in this chapter), the upper layers of the atmosphere, called the *ionosphere,* reflect the waves back to Earth. The reflection of radio waves is called *sky wave propagation.* Depending on the angle at which the signal is reflected, a sky wave path can be as long

as 2,000 miles. HF signals often bounce between the Earth's surface and the ionosphere several times so that contacts are made worldwide. At VHF, multiple hops are rare, but other reflecting mechanisms are present.

- **Tropospheric propagation:** Apart from the ionosphere, the atmosphere itself can direct radio waves. *Tropospheric propagation,* or *tropo,* occurs along weather fronts, temperature inversions, and other large-scale features in the atmosphere. Tropo is common at frequencies in the VHF and UHF range, often supporting contacts over 1,000 miles or more.
- **Aurora:** When the aurora is strong, it absorbs HF signals but reflects VHF and UHF signals while adding a characteristic rasp or buzz. Hams who are active on those bands know to point their antennas north to see whether the aurora can support an unusual contact.
- **Meteor trails:** Meteor trails are very hot from the friction of the meteor's passage through the atmosphere — so hot that the gases become electrically conductive and reflect signals until they cool. For a few seconds, a radio mirror floats high above the Earth's surface. Meteor showers are popular times to try meteor-scatter mode (see Chapter 11).

Dealing with noise digitally

One limiting factor for all wireless communication is noise. Certainly, trying to use a radio in a noisy environment such as a car presents some challenges, but I'm talking here about electrical noise, created by natural sources such as lightning, the aurora, and even the Sun. Other types of noise are human-made, such as arcs and sparks from machinery and power lines. Even home appliances make noise — lots of it. When noise overpowers the signal, radio communication becomes very difficult.

Radio engineers have been fighting noise since the early days of AM radio. FM was invented and used for broadcasting because of its noise-fighting properties. Even so, there are practical limits to what transmitters and receivers can do, which is where digital technology comes in. By using sophisticated methods of turning speech and data into digital codes, digital technology strips away layers of noise, leaving only the desired signal.

Hams have been in the forefront of applying noise-fighting digital techniques to wireless. The noise-canceling technology in most mobile phones was pioneered in part by Phil Karn, a vice president of technology for Qualcomm and amateur operator KA9Q. Recently, powerful noise-fighting coding and decoding techniques have been applied to amateur signals by Nobel Prize laureate Dr. Joe Taylor, also known by his ham radio call sign K1JT. Using Taylor's special software, known as JT65, hams can communicate with signals hundreds of times weaker than the natural noise level, even bouncing their signals off the Moon with simple equipment. JT65 is part of Taylor's WSJT software package, which you can download for free at `http://physics.princeton.edu/pulsar/K1JT`.

Ham radios, CB radios, and mobile phones

Radios abound — enough to boggle your mind. Here are the differences between your ham radio and those other radios:

- **Citizens Band (CB):** CB radio uses 40 channels near the 28 MHz ham band. CB radios are low-power and useful for local communications only, although the radio waves sometimes travel long distances. You don't need a license to operate a CB radio. This lightly regulated service is plagued by illegal operation, which diminishes its usefulness.
- **Family Radio Service (FRS) and General Mobile Radio Service (GMRS):** These popular radios, such as the Motorola Talkabout models, are designed for short-range communications between family members. Usually handheld, both types operate with low-power on UHF frequencies. FRS operation is unlicensed, but using the GMRS channels (see the radio's operating manual or guide) does require a license.
- **Broadcasting:** Although hams are often said to be broadcasting, this term is incorrect. Hams are barred from doing any one-way broadcasting of programs the way that AM, FM, and TV stations do. Broadcasting without the appropriate license attracts a lot of attention from a certain government agency whose initials are FCC.
- **Public-safety and commercial mobile radio:** The handheld and mobile radios used by police officers, firefighters, construction workers, and delivery-company couriers are similar in many ways to VHF and UHF ham radios. In fact, the frequency allocations are so similar that hams often convert and use surplus equipment. Commercial and public-safety radios require a license to operate.
- **Mobile telephones:** Obviously, you don't need a license to use a mobile phone, but you can communicate only through a licensed service provider on one of the mobile phone allocations near 2 GHz. Although the phones are actually small UHF and microwave radios, they generally don't communicate with other phones directly and are completely dependent on the mobile phone network to operate.

Chapter 3

Finding Other Hams: Your Support Group

In This Chapter

- Finding a club
- Becoming a member of the ARRL
- Finding a specialty organization
- Checking out online communities
- Going to hamfests and conventions

One of the oldest traditions of ham radio is helping newcomers. Hams are great at providing a little guidance or assistance. You can make your first forays into ham radio operating much easier and more successful by taking advantage of those helping hands. This chapter shows you how to find them.

Joining Radio Clubs

The easiest way to get in touch with other hams is through your local radio club. The following points hold true for most hams and clubs:

- Most hams belong to a general-interest club as well as one or two specialty groups.
- Most local or regional clubs have in-person meetings, because membership is drawn largely from a single area.
- Almost all clubs have a website or social media site and an e-mail distribution list.
- Specialty clubs focus on activities. Activities such as contesting, low-power operating, and amateur television may have a much wider (even international) membership. See "Taking Part in Specialty Groups," later in this chapter, for more information.

Checking out a club

I found this listing for one of the largest clubs in western Washington state through the ARRL website:

Mike & Key Amateur Radio Club

City: Renton, WA

Call Sign: K7LED

Specialties: Contest, Digital Modes, DX, General Interest, Public Service/Emergency, Repeaters, School or Youth Group, VHF/UHF

Services Offered: Club Newsletter, Entry-Level License Classes, General or Higher License Classes, Hamfest, License Test Sessions, Mentor, Repeater

Section: WWA

Links: www.mikeandkey.org

This club is well suited to a new ham. You'll find yourself in the company of other new license holders, so you won't feel self-conscious about asking questions. The club offers educational programs, activities, and opportunities for you to contribute.

Clubs are great resources for assistance and mentorship (see "Finding Mentors," later in this chapter). As you get started in ham radio, you'll find that you need answers to a lot of basic questions. I recommend that you start by joining a general-interest club (see the next section). If you can find one that emphasizes assistance to new hams, so much the better. You'll find the road to enjoying ham radio a lot smoother in the company of others, and you'll find other new hams to share the experience.

Finding and choosing a club

Here's one way to find ham radio clubs in your area:

1. **Go to** www.qrz.com/clubs**.**
2. **Select your state to find a list of radio-club websites.**
3. **Enter your city or zip code to locate nearby clubs.**

   For an example club listing, see the nearby sidebar "Checking out a club."

The ARRL, covered later in this chapter, also has a directory of affiliated clubs at www.arrl.org/find-a-club.

If more than one club is available in your area, how do you make a choice? Consider these points when making a decision:

- **Which club has meetings that are most convenient for you?** Check out the meeting times and places for each club.
- **Which club includes activities or programs that include your interests?** If a club has a website or newsletter, review the past few month's activities and programs to see whether they sound interesting.
- **Which club feels most comfortable to you?** Don't be afraid to attend a meeting or two to find out what different clubs are like.

You'll quickly find out that the problem isn't finding clubs, but choosing among them. Unless a club has a strong personal-participation aspect, such as a public-service club, you can join as many as you want just to find out about that part of ham radio.

Participating in meetings

After you pick a general-interest club, show up for meetings, and make a few friends right away, your next step is to start participating. But how?

Obviously, you won't start your ham club career by running for president at your second meeting, but ham clubs are pretty much like all other hobby groups, so you can become an insider by following a few easy first steps. You're the new guy or gal, which means you have to show that you want to belong. Here are some ways to assimilate:

- Show up early to help set up, make coffee, hang the club banner, help figure out the projector, and so on. Stay late and help clean up, too.
- Be sure to sign in, sign on, or sign up if you have an opportunity to do so, especially at your first meeting.
- Wear a name tag or other identification that announces your name and call sign in easy-to-read letters.
- Introduce yourself to whomever you sit next to.
- Introduce yourself to a club officer as a visitor or new member. If a "stand up and identify yourself" routine occurs at the beginning of the meeting, be sure to identify yourself as a new member or visitor. If other people also identify themselves as new, introduce yourself to them later.
- After you've been to two or three meetings, you'll probably know a little about some of the club's committees and activities. If one of them

sounds interesting, introduce yourself to whoever spoke about it and offer to help.

- Show up at as many club activities and work parties as possible.
- Comb your hair. Brush your teeth. Sit up straight. Wear matching socks. (Yes, Mom!)

These magic tips aren't just for ham radio clubs; they're for just about any club. Like all clubs, ham clubs have their own personalities, varying from wildly welcoming to tightly knit, seemingly impenetrable groups. After you break the ice with them, though, hams seem to bond for life.

When you're a club elder yourself, be sure to extend a hand to new members. They'll appreciate it just as much as you did when you were in their shoes.

Getting more involved

Now that you're a regular, how can you get more involved? This section gives you some pointers.

Volunteering your services

In just about every ham club, someone always needs help with the following events and activities:

- **Field Day:** Planners and organizers can always use a hand with getting ready for this June operating event (see Chapter 1). Offer to help with generators, tents, and food, and find out about everything else as you go.

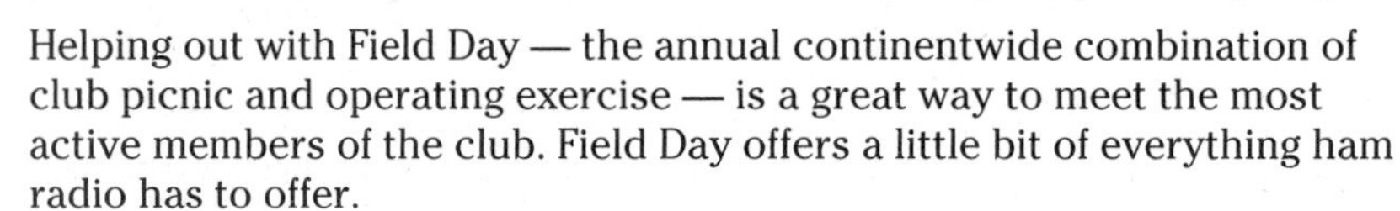

  Helping out with Field Day — the annual continentwide combination of club picnic and operating exercise — is a great way to meet the most active members of the club. Field Day offers a little bit of everything ham radio has to offer.

- **Conventions or hamfests:** If the club hosts a regular event, its organizers probably need almost every kind of help. If you have any organizational or management expertise, so much the better. (I discuss hamfests and conventions in detail later in this chapter.)
- **Awards and club insignia:** Managing sales of club insignia is a great job for a new member. You can keep records, take orders, and make sales at club meetings.

  If you have a flair for arts or crafts, don't be afraid to make suggestions about designing these items.

- **Libraries and equipment:** Many clubs maintain a library of reference books or loaner equipment. All you have to do is keep track of everything and make it available to other members.
- **Club station:** If your club is fortunate enough to have its own radio shack or repeater station, somebody always needs to do maintenance work, such as working on antennas, changing batteries, tuning and testing radios, or just cleaning. Buddy up with the station manager, and you can become familiar with the equipment very quickly. You don't have to be technical — just willing.
- **Website:** If you can write or work on websites, don't hesitate to volunteer your services to the club newsletter editor or webmaster. Chances are that this person has several projects backlogged and would be delighted to have your help.

Find out who's currently in charge of all these areas, and offer your help. You'll discover a new aspect of ham radio, gain a friend, and make a contribution.

Taking part in activities

Along with holding ongoing committee meetings and other business, most clubs sponsor several activities throughout the year. Some clubs are organized around one major activity; others seem to have one or two going on every month. Here are a few common club activities:

- **Public service:** This activity usually entails providing communication services during a local sporting or civic event, such as a parade or festival. Events like these are great ways for you to hone your operating skills.
- **Contests and challenges:** Operating events are great fun, and many clubs enter on-the-air contests as a team or club. Sometimes, clubs challenge each other to see which can generate the most points. You can either get on the air yourself or join a multiple-operator station. (For more on contests, see Chapter 11.)
- **Work parties:** What's a club for if not to help its own members? Raising a tower or doing antenna work at the club's station or that of another member is a great way to meet active hams and discover this important aspect of station building.
- **Construction projects:** Building your own equipment and antennas is a lot of fun, so clubs occasionally sponsor group construction projects in which everyone builds a particular item at the same time. Building your own equipment saves money and lets everyone work together to solve problems. If you like building things or have technical skills, taking part in construction projects is a great way to help out.

Supporting your club by participating in activities and committees is important. For one thing, you can acknowledge the help you get from the other members. You also start to become a mentor to other new members. By being active within the club, you strengthen the organization, your friendships with others, and the hobby in general.

Exploring ARRL

The *American Radio Relay League* (ARRL; www.arrl.org) is the oldest continuously functioning amateur radio organization in the world. Founded before World War I, it provides services to hams around the world and plays a key part in representing the ham radio cause to the public and governments. That ham radio could survive for nearly 100 years without a strong leadership organization is hard to imagine, and ARRL has filled that role. I devote a whole section of this chapter to ARRL simply because it's such a large presence within the hobby for U.S. hams (and for those in Canada who belong to its sister organization, Radio Amateurs of Canada).

ARRL is a volunteer-based, membership-oriented organization. Rest assured that even as a new ham, you can make a meaningful contribution as a volunteer. To find out how to join, go to www.arrl.org/membership.

ARRL's benefits to you

The most visible benefit of ARRL membership is that you receive *QST* magazine in print or digital format every month (see Figure 3-1). The largest, oldest, and most widely read ham radio magazine, *QST* includes feature articles on technical and operating topics, reports on regulatory information affecting the hobby, the results of ARRL-sponsored competitions, and columns on a wide variety of topics.

Along with the print magazine, ARRL maintains an active and substantial website, providing current news and general-interest stories; the Technical Information Service, which allows you to search technical documents and articles online; and several free e-mail bulletins and online newsletters.

ARRL also manages the Amateur Radio Emergency Service (ARES), which helps hams organize at the local level. ARES teams support local government and public-safety functions with emergency communication services. They also perform public service by providing support and communications services for parades, sporting events, and similar events. You can find out more about ARES in Chapter 10.

Figure 3-1: *QST* covers most aspects of ham radio every month.

Courtesy American Radio Relay League

In addition, ARRL is the largest single sponsor of operating activities for hams, offering numerous contests, award programs, and technical and emergency exercises.

ARRL's benefits to the hobby

By far the most visible aspect of ARRL on the ham bands is its headquarters station, W1AW (see Figure 3-2). Carrying the call sign of ARRL founder Hiram Percy Maxim, the powerful station beams bulletins and Morse code practice sessions to hams around the planet every day. Visiting hams can even operate the W1AW station themselves (as long as they remember to bring a license). Most hams think that being at the controls of one of the most famous and storied ham stations in the world is the thrill of a lifetime.

Figure 3-2: W1AW in Newington, Connecticut.

ARRL is a Volunteer Examiner Coordinator (VEC) organization, and you may have taken your licensing test at an ARRL-VEC exam session. With the largest number of Volunteer Examiners (VEs), the ARRL-VEC helps thousands of new and active hams take their licensing exams, obtain vanity and special call signs, renew their licenses, and update their license information of record. When the Federal Communications Commission (FCC) could no longer maintain the staff to administer licensing programs, ARRL and other ham organizations stepped forward to create the largely self-regulated VEC programs that are instrumental to healthy ham radio.

One of the least visible of ARRL's functions, but arguably one of its most important, is its advocacy of amateur radio service to governments and regulatory bodies. In this telecommunications-driven age, the radio spectrum is valuable territory, and many commercial services would like to get access to amateur frequencies, regardless of the long-term effects. ARRL helps regulators and legislators understand the special nature and needs of amateur radio.

ARRL's benefits to the public

Although it naturally focuses on its members, ARRL takes its mission to promote amateur radio seriously. To that end, its website is largely open to the public, as are all bulletins broadcast by W1AW (see the preceding section). The organization also provides these services:

- **Facilitates emergency communications:** In conjunction with the field organization, ARES teams around the country provide thousands of hours of public service every year. While individual amateurs render valuable aid in times of emergency, the organization of these efforts multiplies the usefulness of that aid. ARRL staff members also help coordinate disaster response across the country.
- **Publishes the *ARRL Handbook for Radio Communications:*** Now in its 90th edition, the handbook (`www.arrl.org/shop`) is used by telecommunications professionals and amateurs alike.
- **Provides technical references:** The league publishes numerous technical references and guides, including conference proceedings and standards.
- **Promotes technical awareness and education:** ARRL is involved with the Boy Scouts' and Girl Scouts' Radio merit badge and with Jamboree-on-the-Air programs. It also sponsors the Teachers Institute on Wireless Technology to train and license primary and secondary educators.

Taking Part in Specialty Groups

Ham radio is big, wide, and deep. The hobby has many communities that fill the airwaves with diverse activities. A *specialty club or organization* focuses on one aspect of ham radio that emphasizes certain technologies or types of operation. Many specialty organizations have worldwide membership.

Some clubs focus on particular operating interests, such as qualifying for awards or operating on a single band. An example of the latter is the 10-10 International Club (`www.ten-ten.org`), which is for operators who prefer the 10 meter band — a favorite of low-power and mobile stations. The 10-10 club sponsors several contests every year and offers a set of awards for contacting its members. A similar group, the Six Meter International Radio Klub (SMIRK), promotes activity on the 6 meter band, including unusual methods of signal propagation. You can find information about the club's contests and awards at `www.smirk.org`.

To find specialty clubs, search your favorite search engine for your area of interest and the phrase *radio club*. Using the search term *10 meter amateur radio club,* for example, turns up a bunch of ham clubs and forums about operating on the 10 meter band.

This section lists only a few of the specialized groups you'll find in ham radio; there are many, many more.

Competitive clubs

One type of specialty club is the contest club. Members enjoy participating in competitive on-the-air events known as contests or radiosport (see Chapter 11). These clubs challenge one another, sponsor awards and plaques, and generally encourage their members to build up their stations and techniques to become top contest operators.

Contest clubs tend to be local or regional due to the rules of club competition. You can view an extensive list of clubs that compete in the ARRL club competition at `www.arrl.org/contest-club-list`.

No less competitive than contest operators are the long-distance communications specialists, or *DXers,* who specialize in contacts with places well off the beaten track. The quest to *work 'em all* (contact every country on every ham band) lasts a lifetime, so DXers form clubs to share operating experiences and host traveling hams, fostering international communications and goodwill along the way.

Many contesters are also DXers, and vice versa. Because of the international nature of DXing and contesting, clubs that specialize in these activities tend to have members sprinkled around the globe. You can find lists of these organizations at `www.dailydx.com/clubs.htm`.

Handiham

Ham radio provides excellent communication opportunities to people who otherwise find themselves constrained by physical limitations. Handiham (`www.handiham.org`), founded in 1967, is dedicated to providing tools that make ham radio accessible to people with disabilities of all sorts, helping them turn their disabilities into assets. The website provides links to an extensive set of resources.

Handiham not only helps hams with disabilities reach out to the rest of the world, but also helps its members link up with other members and helpful services.

Even if you're not disabled, Handiham may be a welcome referral for someone you know, or you may want to volunteer your services.

The CQ Communications family of print and digital magazines (see Figure 3-3) provides a lot of good information on ham specialties. *CQ* focuses on general-interest stories and news, product reviews, and columns on technical and operating interests. *CQ VHF* is designed for hams operating at 50 MHz and above — which includes most new hams! Two additional magazines — *World Radio Online* and *Popular Communications* — round out the selections.

AMSAT

AMSAT (short for Amateur Radio SATellite Corporation, `www.amsat.org`) is an international organization that helps coordinate satellite launches and oversees the construction of its own satellites. Yes, Virginia, there really are amateur radio satellites whizzing through the heavens! The first one, launched in 1962, sent a Morse code beacon consisting of the letters *HI* (in Morse code speak, "di-di-di-dit, di-dit"), known as "the telegrapher's laugh." The first, OSCAR-1 (Orbiting Satellite Carrying Amateur Radio), was about the size of a briefcase.

These days, many satellites either provide communication relay services to hams or use ham radio for telemetry and beacons in support of academic science experiments. In 2011, for example, University of Texas students built and launched a pair of nano-satellites known as FASTRAC (see Figure 3-4) to study orbital maneuvering.

Figure 3-3: CQ Communications publications cover just about every style and interest in ham radio.

Magazine covers courtesy CQ Communications, Inc.

Figure 3-4: University of Texas students built amateur satellite FASTRAC.

Courtesy American Radio Relay League

Radio operation via satellite is a lot easier than you may think, however, as you can find out in Part IV of this book. All you need to make contacts through — or with — satellites is some simple equipment. Figure 3-5 shows Sean Kutzko (call sign KX9X) using a handheld radio and a hand-aimed antenna to make contacts through satellite AO-27.

Figure 3-5: Sean Kutzko (KX9X) shows that satellite operation can be easy!

TAPR

Tucson Amateur Packet Radio (TAPR; www.tapr.org) has been instrumental in bringing modern digital communications technology to ham radio. In return, TAPR members created several innovative communication technologies that are now commonplace beyond ham radio, such as the communications system known as *packet radio,* which is widely used in industry and public safety. Recently, TAPR members have been involved in modern digital communications technology by developing software-defined radio (SDR) components, as shown in Figure 3-6.

Figure 3-6: TAPR supports the development of high-performance SDR technology and special protocols to send voice and data via ham radio.

Courtesy Phil Harman (VK6KPH) and Tucson Amateur Packet Radio, Inc.

If you have a strong computer or digital technology background, TAPR is likely to have activities that pique your interest.

YLRL

The Young Ladies' Radio League (YLRL; www.ylrl.org) is dedicated to promoting ham radio to women, encouraging them to be active on the air, promoting women's interests within the hobby, and providing a membership organization for female hams.

The organization has chapters in many countries, some of which host conventions, thereby creating opportunities for members to travel.

The YLRL's website provides a list of activities and member services. The organization also has a vigorous awards program; it sponsors on-the-air nets and on-the-air competitions for members throughout the year.

QRP clubs

QRP is ham radio shorthand for *low-power operating,* in which hams use just a few watts of power to span the oceans. Like bicyclists among motorists, QRP enthusiasts emphasize skill and technique, preferring to communicate by using minimal power. They're among the most active designers and builders of any group in ham radio. If you like building your own gear and operating with a minimum of power, check out these clubs and other groups of QRPers.

One way to find QRP clubs is to visit www.arrl.org/find-a-club and search for *QRP.*

The largest U.S. QRP club is QRP Amateur Radio Club International, known as QRP ARCI (www.qrparci.org). Its magazine, *QRP Quarterly* (see Figure 3-7), is full of construction projects and operating tips.

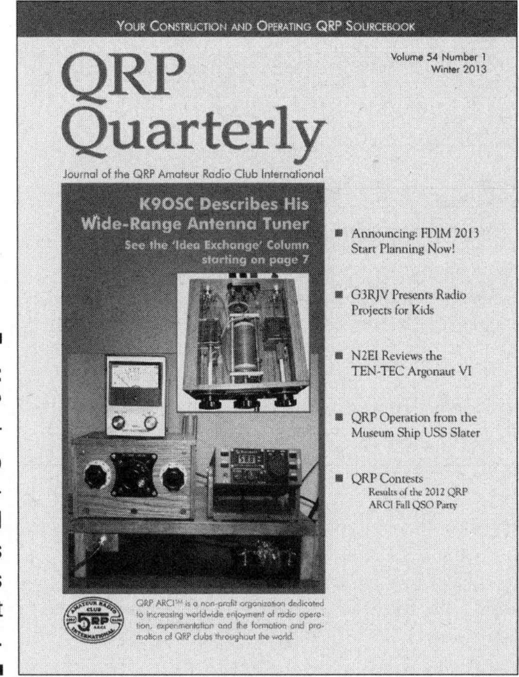

Figure 3-7: The QRP Amateur Radio Club International publishes this excellent quarterly.

Courtesy QRP ARCI

Many QRP clubs have worldwide membership. One of my favorites is the British club GQRP. (*G* is a call sign prefix used by stations in England.) You can find the GQRP Club website at www.gqrp.com.

Interacting in Online Communities

Just like every other human activity, ham radio has online communities in which members discuss the various aspects of the hobby, provide resources, and offer support 24 hours a day. Will these communities replace ham radio? Not likely; the magic of radio is too strong. By their presence, though, they make ham radio stronger by distributing information, cementing relationships, and adding structure.

Social media

Everything has a presence online, and ham radio is no different. Just search for *ham radio* on Facebook, for example, and you'll find dozens of possibilities, ranging from general-interest clubs to emergency communications to license-exam practice to contesting — and more.

Here are links to a few popular streams of information about amateur radio:

- **Twitter:** https://twitter.com/amateurradio (@Amateurradio) and https://twitter.com/hamradio (@hamradio)
- **Reddit:** www.reddit.com/r/amateurradio and www.reddit.com/r/hamradio

E-mail reflectors

The first online communities for hams were e-mail lists, known as *reflectors.* Reflectors are mailing lists that take e-mail from one mailbox and rebroadcast it to all members. With some list memberships numbering in the thousands, reflectors get information spread around pretty rapidly. Every ham radio interest has a reflector.

Table 3-1 lists several of the largest websites that serve as hosts for reflectors. You can browse the directories and decide which list suits your interests. (Be careful, though, that you don't wind up spending all your time on the reflectors and none on the air.)

Table 3-1 Hosts and Directories for Ham Radio Reflectors

Website	*Topics*
www.qth.com	Radios, bands, operating, and awards
www.contesting.com	TowerTalk, CQ-Contest, Amps, Top Band (160 meters), RTTY (digital modes)
www.dxzone.com/catalog/Internet_and_Radio/Mailing_Lists and www.ac6v.com/mail.htm	Directories of reflectors and forums hosted on other sites

Because my main interests are QRP operating on the HF bands, contesting, and making DX (long-distance) contacts, for example, I subscribe to the QRP reflector, the CQ-Contest reflector, a couple of the DX reflectors, and the Top Band reflector about 160 meter operating techniques and antennas. To make things a little easier on my e-mail inbox, I subscribe in digest format so that I get one or two bundles of e-mail every day instead of many individual messages. Most reflectors are lightly moderated and usually closed to any posts that aren't from subscribed members — in other words, spam.

The reflectors at Yahoo! Groups (http://groups.yahoo.com) and Google Groups (https://groups.google.com) offer a little more than just e-mail distribution. They also offer file storage, a photo-display function, chat rooms, polls, and excellent member management. To take advantage of these services, create a personal Yahoo! profile or Google account; then search the service for *amateur radio* or *ham radio groups*. More than 1,400 ham radio groups are running Yahoo! Groups, for example. You'll find similar numbers on the Google Groups site.

As soon as you settle into an on-the-air routine, subscribe to one or two reflectors. Reflectors are great ways to find out about new equipment and techniques before you take the plunge and try them yourself.

E-mail portals

Portals provide a comprehensive set of services and function as ham radio home pages. They feature news, informative articles, radio buy-and-sell pages, links to databases, reflectors, and many other useful services to hams. The best-known portals are eHam.net (www.eham.net) and QRZ.com (www.qrz.com).

QRZ (the ham radio abbreviation for "Who is calling me?") evolved from a call-sign lookup service — what used to be a printed book known as a *call-book* — to the comprehensive site that you see today. The call-sign search features are incredibly useful, and the site offers a variety of call-sign management functions.

eHam.net is organized around community functions, operating functions, and resources. You can find real-time links to a DX-station spotting system (frequencies of distant stations that are currently on the air) and the latest solar and ionospheric data that affects radio propagation.

As with all public websites, not everyone behaves perfectly, but I recommend that you bookmark both of these sites, which offer lively collections of news and articles along with useful forums and features. Both sites are gateways to e-mail reflectors (see the preceding section) that make subscription easy.

Attending Hamfests and Conventions

Depending on how much you like collecting and bargaining, I may have saved the best for (almost) last. Even in these days of online auction sites, *hamfests* — ham radio flea markets — continue to be some of the most interesting events in ham radio. Imagine an Oriental bazaar crammed with technological artifacts spanning nearly a century, old and new, small and massive, tubes, transistors, computers, antennas, batteries . . . I'm worn out just thinking about it. (I love a good hamfest; can you tell?)

Ham radio conventions have a much broader slate of activities than hamfests do; they may include seminars, speakers, licensing test sessions, and demonstrations of new gear. Some conventions host competitive activities such as foxhunts or direction finding, or they may include a swap meet along with the rest of the functions. Conventions usually have a theme, such as emergency operations, QRP, or digital radio transmissions.

Finding hamfests

In the United States, the best place to find hamfests is ARRL's Hamfests and Conventions Calendar (`www.arrl.org/hamfests-and-conventions-calendar`). Search for events by location or ARRL section or division. The calendar usually lists about 100 hamfests. Most metropolitan areas have several good-size hamfests every year, even in the dead of winter.

Prepping for a hamfest

After you have a hamfest in your sights, set your alarm for early morning, and get ready to be there at the opening bell. Although most are Saturday-only events, more and more are opening on Friday afternoon.

Be sure to bring the following things:

- **An admission ticket:** You need a ticket, sold at the gate or by advance order through a website or e-mail.
- **Money:** Take cash, because most vendors don't take checks or credit cards.
- **Something to carry your purchases in:** Take along a sturdy cloth sack, backpack, or another type of bag that can tolerate a little grime or dust.
- **A handheld or mobile rig:** Most hamfests have a talk-in frequency, which is almost always a VHF or UHF repeater. If you're unfamiliar with the area and don't have a GPS unit to guide you, get directions while you're en route.

  If you attend with a friend, and both of you take handheld radios, you can share tips about the stuff you find while walking the aisles.
- **Water and food:** Don't count on food being available, but the largest hamfests almost always have a hamburger stand. Gourmet food is rarely on hand; expect the same level of quality that you'd find at a ballpark concession stand. Taking along a full water bottle is a good idea.

Buying at hamfests

After parking, waiting, and shuffling along in line, you finally make it inside the gates, and you're ready to bargain. No two hamfests are alike, of course, but here are some general guidelines to live by, particularly for hamfest newcomers:

- If you're new to ham radio, buddy up with a more experienced ham who can steer you around hamfest pitfalls.
- Most prices are negotiable, especially after lunch on Saturday, but good deals go quickly.
- Most vendors aren't interested in trades, but you do no harm by offering.
- Hamfests are good places to buy accessories for your radio, often for a fraction of the manufacturer's price if they're sold separately from the radio. Commercial vendors of new batteries often have good deals on spare battery packs.

- Many hamfests have electricity available so that vendors can demonstrate equipment and maybe even a radio test bench. If a seller refuses to demonstrate a supposedly functional piece of gear or won't open a piece of equipment for inspection, you may want to move along.

- Unless you really know what you're doing, avoid antique radios. They often have quirks that can make using them a pain or that require impossible-to-get repair parts.
- Don't be afraid to ask what something is. Most of the time, the ham behind the table enjoys telling you about his or her wares, and even if you don't buy anything, the discussion may attract a buyer.

- Be familiar with the smell of burned or overheated electronics, especially transformers and sealed components. Direct replacements may be difficult to obtain.
- If you know exactly what you're looking for, check auction and radio swap sites such as `www.eham.net` and `www.qrz.com` before and even while attending the hamfest if you have a smartphone. You can get an idea of the going price and average condition, so you're less likely to get gouged.

- The commercial vendors will sell you accessories, tools, and parts on the spot, which saves you shipping charges.
- Don't forget to look *under* the tables, where you can occasionally find some real treasures.

Finding conventions

Conventions tend to be more extravagant affairs, held in hotels or conventions centers, that are advertised in ham radio magazines as well as online. The main purposes are programs, speakers, and socializing.

The two largest ham radio conventions are the Dayton Hamvention (`www.hamvention.com`), held in Ohio in mid-May, and the Internationale Exhibition for Radio Amateurs (`www.hamradio-friedrichshafen.de/ham-en`), held in Friedrichshafen, Germany, in late June. Dayton regularly draws 20,000 or more hams; Friedrichshafen, nearly that many. Both events have mammoth flea markets, an astounding array of programs, internationally known speakers, and more displays than you can possibly see.

ARRL national and division conventions (listed on the ARRL website at `www.arrl.org/hamfests-and-conventions-calendar`) are held in every region of the United States. Radio Amateurs of Canada (`www.rac.ca`) also hosts a national convention every year. These conventions typically attract a few hundred to a few thousand people and are designed to be family friendly.

They also provide a venue for specialty groups to host conferences within the overall event. These smaller conferences offer extensive programs on regional disaster and emergency communications, direction finding, QRP, county hunting, wireless networking on ham bands, and so on.

Some conventions and conferences emphasize one of ham radio's many facets, such as DXing, VHF and UHF operating, or digital technology. If you're a fan of a certain mode or activity, treating yourself to a weekend convention is a great way to meet hams who share your tastes and to discover more about your interests. Table 3-2 lists a few of the specialty conventions held around the United States each year.

Table 3-2 Specialty Conventions

Name	*Theme*	*Website*
International DX Convention (hosted alternately by the Northern and Southern California Contest Clubs)	DX and contesting	`www.dxconvention.com`
W9DXCC (hosted by the Northern Illinois DX Association)	DX and contesting	`www.w9dxcc.com`

Pacific Northwest DX Convention (rotates among clubs in Seattle, Portland, Spokane, and Vancouver)	DX and contesting	`www.pnwdx.org`
		`www.wvdxc.org`
		`www.wwdxc.org`
		`www.sdxa.org`
		`www.orcadxcc.org`

SVHFS Conference (hosted by the Southeastern VHF Society)	VHF, UHF, and microwaves	`www.svhfs.org`
International EME Conference	EME (Earth–Moon–Earth) operating	`www.eme2012.com` (or search for *International EME Conference* online)
Digital Communications Conference (hosted by ARRL and TAPR)	Digital communications	`www.tapr.org/conferences.html`

Finding Mentors

A mentor is very useful in helping you over the rough spots that every newcomer encounters. A good place to start your search for a mentor is to search for ham radio clubs in your area (refer to "Finding and choosing a club," earlier in this chapter). You might start on the clubs page of the QRZ.com website (`www.qrz.com/clubs`), for example. When you've narrowed down the clubs closest to you, enter **mentor** in the Tag window to find clubs that offer special help to new hams.

As your interests widen, you'll need additional help. Luckily, hundreds of potential mentors, known as *Elmers* (see the next section and Chapter 5), are available in ham radio organizations around the world.

Getting a hand from Elmer

As a new ham, you may want to join one of the Elmer e-mail lists that are set up specifically to answer questions and offer help. To find general and topical Elmer lists, enter **ham radio elmer reflector** in a search engine, and you'll turn up several candidates.

Finding help online

There's nothing quite like a demonstration to find out how to do something, such as put on a connector, make a contact, tune an antenna, or assemble a kit. Many online video and photography websites can speed you on your way to ham radio success; YouTube (`www.youtube.com`) and Instructables (`www.instructables.com`) are just two of the options.

Also available are several nicely produced talk show–style programs that have large followings. Here are a few of my favorites:

- *Ham Nation* on TWiT.TV (`http://twit.tv/shows`), covers operating and technical topics in an informal and fast-paced format. A new show airs every week.
- *Ham Radio Now TV,* hosted by Gary Pearce (KN4AQ), is a weekly podcast that tackles all sorts of interesting topics. Pearce's web page (`www.arvideonews.com/hrn`) lists many other audio and video programs.
- One of my favorite technical programs is *SolderSmoke* (`www.soldersmoke.com`), a monthly audio program that covers topics associated with building and repairing your own equipment. It's great to download and listen to in the car.

Finally, webinars (online video seminars hosted by an instructor) are becoming common as broadband Internet access becomes the norm. Many of these events are archived, such as those hosted by the World Wide Radio Operator's Foundation (`http://wwrof.org`). A webinar is the next-best thing to your mentor's being there in the room with you.

Part II

Wading through the Licensing Process

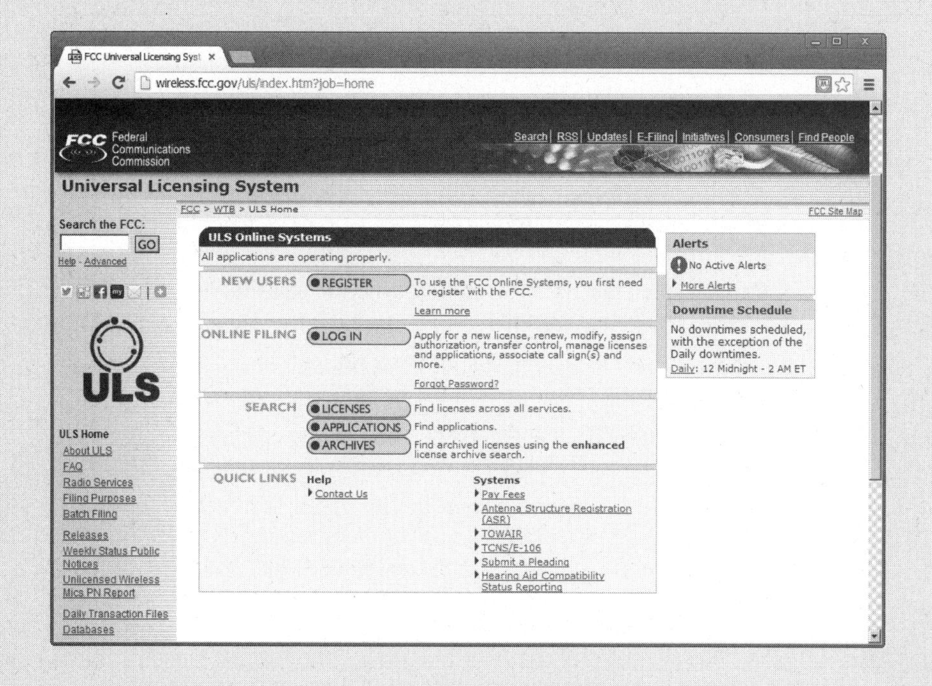

The Web Extras site for this book (at www.dummies.com/extras/hamradio) has copies of the forms you'll encounter when you get ready to take the test and some sample questions from the Technician class exam to show you what they look like and how to choose an answer.

In this part . . .

- Tour the Federal Communications Commission's amateur radio licensing system, see how call signs are structured, and find out how the call sign system works.
- Find out about the test itself, as well as how to study for it by yourself and with a mentor.
- See how to locate and register for a test session, and get ready to take the test.
- While you're waiting for your brand-new call sign, decide whether you want to customize your call.
- The Web Extras site for this book (at `www.dummies.com/extras/hamradio`) has copies of the forms you'll encounter when you get ready to take the test and some sample questions from the Technician class exam to show you what they look like and how to choose an answer.

Chapter 4

Figuring Out the Licensing System

In This Chapter

- Finding out about the amateur service
- Understanding license types
- Studying for your license exam
- Testing for your license
- Comprehending your call sign

Unlike some of the other types of radios available to the public, ham radios can't be used without a license. Like most people, you're probably familiar with the process of getting a license to drive your car, to fish, or to get married, but ham radio licensing is a little different. The process is easy to deal with when you know how it works, however.

Amateur radio is one of many types of *services* that use the radio waves to communicate. Other services include broadcast (AM and FM radio, television), public safety (police and fire departments), aviation, and even radar systems (radio navigation).

When the name of a specific service is capitalized, such as Amateur Radio Service or Citizens Band, that's a formal reference to the set of Federal Communications Commission (FCC) rules for that service. Each service has a different set of rules set up for its type of operating and use.

To maintain order on the airwaves, the FCC requires that each signal come from a licensed station. Stations in all the different services must abide by FCC regulations to obtain and keep their licenses, which give them permission to transmit according to the rules for that service. That's what a ham license is: authority for you to transmit on the frequencies that licensees of the amateur radio service are permitted to use. This chapter explains the FCC licensing system for amateur radio in the United States.

Getting an Overview of the Amateur Service

By international treaty, the amateur service in every country is a licensed service — that is, a government agency has to approve the application of every ham to transmit. Although regulation may seem to be a little quaint, given all the communications gadgets for sale these days, licensing is necessary for a couple of reasons:

- It allows amateurs to communicate internationally and directly without using any kind of intermediate system that regulates their activities.
- Because of the power and scope of amateur radio, hams need a minimum amount of technical and regulatory background so that they can coexist with other radio services, such as broadcasting.

FCC rules

By maintaining the quality of licensees, licensing helps ensure that the amateur service makes the best use of its unique citizen access to the airwaves. Licensing sets ham radio apart from the unlicensed services and is recognized in FCC Rule 97.1:

- Recognition of ham radio's exceptional capability to provide emergency communications (Rule 97.1(a))
- Promote the amateur's proven ability to advance the state of the radio art (Rule 97.1(b))
- Encourage amateurs to improve their technical and communications skills (Rule 97.1(c))
- Expand the number of trained operators, technicians, and electronics experts (Rule 97.1(d))
- Promote the amateur's unique ability to enhance international goodwill (Rule 97.1(e))

Pretty heady stuff, eh? Ham radio does all these good things in exchange for access to radio space along with the commercial and government users. You can find all the pertinent rules at `http://wireless.fcc.gov/index.htm?job=rules_and_regulations`; click the Part 97 link for the amateur radio rules. Plain-English discussion of the rules is available in *FCC Rules and Regulations for the Amateur Radio Service,* published by the American Radio Relay League (ARRL; see Chapter 3).

Frequency allocations

To keep order in the growing radio communications field, a group of countries got together as the International Telecommunication Union (`www.itu.int`), based in Geneva, Switzerland. The ITU, which is part of the United Nations, provides a forum for countries to create and administer rules of radio spectrum use.

The ITU divides the spectrum into small ranges in which specific types of uses occur (see Figure 4-1). These ranges are *frequency allocations,* which hams call *bands.*

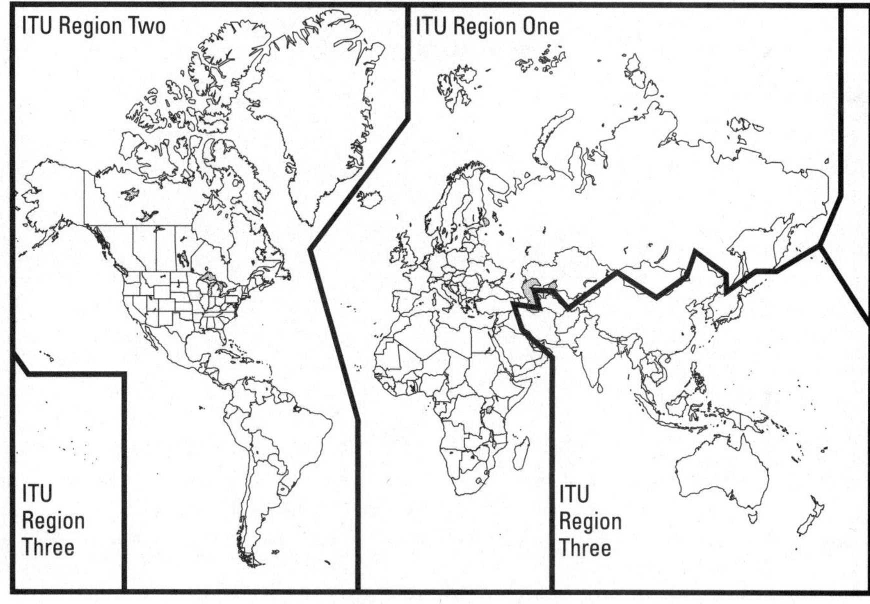

Figure 4-1: ITU region map showing the world's three telecommunications administrative regions.

The world is divided into three regions, as follows:

- **Region 1:** Europe, Russia, and Africa
- **Region 2:** North and South America
- **Region 3:** Asia, Australia, and most of the Pacific

Within each region, each type of radio service — amateur, military, commercial, and government — is allocated a share of the available frequencies. Luckily for amateurs, most of their allocations are the same in all three regions, so they can talk to one another directly.

Figure 4-2 shows the high-frequency (HF) range frequencies (from 3 MHz to 30 MHz). This allocation is very important, particularly on the long-distance bands, where radio signals might propagate all the way around the Earth. Talking to someone in a foreign country is pretty difficult if you can't both use the same frequency.

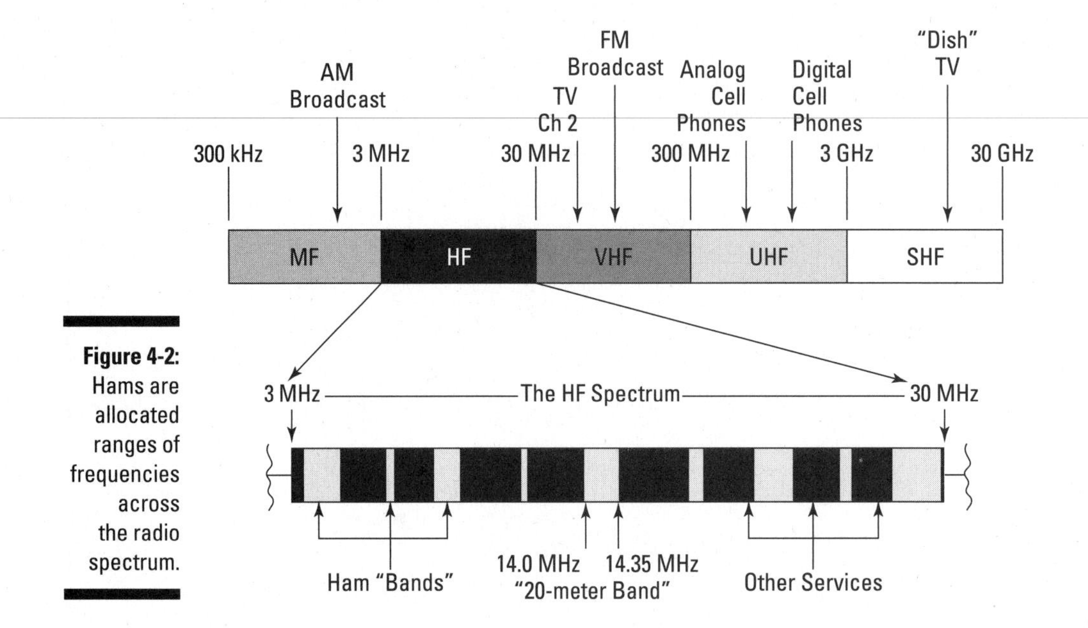

Figure 4-2: Hams are allocated ranges of frequencies across the radio spectrum.

To get an idea of the complexity of the allocations, browse to the Region 2 allocation chart at `www.ntia.doc.gov/files/ntia/publications/2003-allochrt.pdf`. (If you have Adobe Acrobat, you can download the chart in full color.) The individual colors represent different types of radio services. Each service has a small slice of the spectrum, including amateurs. (Can you find the amateur service on the chart? Hint: It's green.)

Amateurs have small allocations at numerous places in the radio spectrum, and access to those frequencies depends on the class of license you hold (see the next section). The higher your license class, the more frequencies you can use.

Choosing a Type of License

Three types of licenses are being granted today: Technician, General, and Amateur Extra.

By taking progressively more challenging exams, you gain access to more frequencies and operating privileges, as shown in Table 4-1. After you pass a specific test level, called an *element,* you have permanent credit for it as long as you keep your license renewed. This system allows you to progress at your own pace. Your license is good for ten years, and you can renew it without taking an exam.

Table 4-1 Privileges by License Class

License Class	*Privileges*	*Notes*
Technician	All amateur privileges above 50 MHz; limited CW, Phone, and Data privileges below 30 MHz	
General	Technician privileges plus most amateur HF privileges	
Amateur Extra	All amateur privileges	Small exclusive sub-bands are added on 80, 40, 20, and 15 meters.

Technician class

Nearly every ham starts with a Technician class license, also known as a Tech license. A Technician licensee is allowed access to all ham bands with frequencies of 50 MHz or higher. These privileges include operation at the maximum legal power limit and using all types of communications. Tech licensees may also transmit using voice on part of the 10 meter band and Morse code on some of the HF bands below 30 MHz.

The test for this license consists of 35 multiple-choice questions on regulations and technical radio topics. You have to get 26 or more correct to pass.

Morse code was once required for amateur operation below 30 MHz because of international treaties adopted when a great deal of commercial and military radio traffic — news, telegrams, ship-to-ship, and ship-to-shore messages — was conducted with the code. Emergency communications were often coded, too. Back then, using Morse code was considered to be a standard radio skill. Morse code still makes up a great deal of amateur operations, from casual ragchewing to passing messages, participating in contests, and providing emergency operations. Its efficient use of transmitted power and spectrum space, as well as its innate musicality and rhythm, make it very popular with hams. Also, it's easy and fun to use. Chapter 9 tells you all about Morse code.

General class

After earning the entry-level Technician license, many hams immediately start getting ready to upgrade to a General class license. When you obtain a General class license, you've reached a great milestone. General class licensees have full privileges on nearly all amateur frequencies, with only small portions of some HF bands remaining off limits.

The General class exam, which includes 35 questions (you have to get 26 right to pass), covers many of the same topics as the Technician exam, but in more detail. The exam introduces some new topics that an experienced ham is expected to understand.

Amateur Extra class

General class licensees still can't access everything; the lowest segments of several HF bands are for Amateur Extra class licensees only. These segments are where the expert Morse code operators hang out and are considered to be prime operating territory. If you become interested in contesting, contacting rare foreign stations (*DXing;* see Chapter 11), or just having access to these choice frequencies, you want to get your Amateur Extra license — the top level.

The Amateur Extra exam consists of 50 multiple-choice questions, 37 of which you must answer correctly to pass. The exam covers additional rules and regulations associated with sophisticated operating and several advanced technical topics. Hams who pass the Amateur Extra exam consider their license to be a real achievement. Do you think you can climb to the top rung of the licensing ladder?

Grandfathered classes

The amateur service licensing rules have changed over the years, reducing the number of license classes. Hams who hold licenses in deleted classes may renew those licenses indefinitely, but no new licenses for those classes are being issued.

Two grandfathered license classes remain:

- **Novice:** The Novice license was introduced in 1951 with a simple 20-question test and 5-words-per-minute code exam. A ham with a General class (or higher) license administered the exam. Originally, the license was good for a single year, at which point the Novice upgraded or had to get off the air. These days, the Novice license, like other licenses, has a ten-year term and is renewable. Novices are restricted to segments of the 3.5, 7, 21, 28, 222, and 1296 MHz amateur bands.

- **Advanced:** Advanced class licensees passed a written exam midway in difficulty between those for the General and Amateur Extra classes. They received frequency privileges between those of General and Amateur Extra licensees.

Table 4-2 shows the relative populations of all type of license holders as of April 2013.

Table 4-2 Relative Populations of License Classes

License Class	*Active Licenses*	*Share of Active Licensees*
Technician	347,580	48.7%
General	164,954	23.1%
Amateur Extra	131,983	18.5%
Advanced	55,714	7.8%
Novice	13,501	1.9%
Total	**713,732**	**100%**

Source: www.ah0a.org/FCC/Licenses.html.

Getting Licensed

To become licensed and receive your "ticket," you'll need to do a little studying, and you have plenty of opportunities to practice. Then you'll take your exam, administered by volunteer hams who were all in your shoes once upon a time. After you pass, you'll receive a call sign that is yours and yours alone: your radio name. Ready? Let's go!

Studying for your exam

ARRL (www.arrl.org) and other organizations publish study guides and manuals, some of which may be available through your local library. Also, online tests are available, listing the actual questions that are on the test (see Chapter 5). Take advantage of these materials, and you'll be confident that you're ready to pass the exam on test day.

The pool of questions for the examinations changes every four years. Make sure that you have the current version of study materials, containing the correct questions and any recent changes in rules and regulations.

I cover the study process in Chapter 5.

Taking the licensing test

In the Olden Days, hams took their licensing tests at the nearest FCC office, which could be hundreds of miles away. I vividly remember making long drives to a government office building to take my exams along with dozens of other hams.

Nowadays, although the FCC still grants the licenses, it no longer administers amateur radio licensing examinations. In the United States, ham radio license exams are given by Volunteer Examiners (VEs); some VEs even file the results with the FCC. This process enables you to get your license and call sign much faster than in the days when the FCC handled everything.

The tests are usually available a short drive away at a club, a school, or even a private home. As of early 2013, it costs $15 to attend a test session and take an exam for any of the license elements.

See Chapter 6 for full details on taking a licensing test.

Volunteer Examiner Coordinators

A *Volunteer Examiner Coordinator* (VEC) organization takes responsibility for certifying and coordinating the Volunteer Examiners (VEs) who run the license exam sessions; it also processes FCC-required paperwork generated by the VEs. Each VEC maintains a list of VEs, upcoming test sessions, and other resources for ham test-takers. It can also help you renew your license and change your address or name.

The VEC with the most VEs is the group run by the American Radio Relay League (ARRL-VEC), but 13 other VECs are located around the United States. Some VECs, such as ARRL-VEC and W5YI-VEC, operate nationwide; others work in only a single region.

You can find a VEC near you at `www.ncvec.org`.

Volunteer Examiners

VEs make the system run. Each exam requires three VEs to be present and to sign off on the test paperwork. VEs are responsible for all aspects of the testing process, including providing the meeting space and announcing the test sessions.

VEs are authorized to administer *(proctor)* license exams for the same class of license they hold themselves or for lower classes. A General class VE, for example, can administer Technician and General exams but not Amateur Extra exams. If they incur any expenses, such as for supplies or facility rental, they're allowed to keep up to $7 per person of the $15 test fee, if they want; the remainder goes to the VEC to cover its expenses.

General, Advanced, and Amateur Extra class licensees can become VEs by contacting one of the VEC organizations and completing whatever qualification process the VEC requires. The ARRL-VEC, for example, provides a booklet on the volunteer licensing system and requires applicants to pass a short exam. VE certification is permanent as long as it is renewed on time with the VEC.

VEs are amateurs just like you; they do a real service to the amateur community by making the licensing system run smoothly and efficiently. Don't forget to say "Thanks!" at the conclusion of your test session, pass or fail. Better yet, become a VE yourself. It's fun and rewarding. As a VE, I've given more than 30 exams to hams as young as 10 years old.

Understanding Your Call Sign

Each license that the FCC grants comes with a very special thing: a unique call sign (*call,* to hams). Your call sign is both certification that you have passed the licensing exam and permission to construct and operate a station — a special privilege. If you're a new licensee, you'll get your call sign within seven to ten business days of taking your licensing exam.

Your call sign becomes your on-the-air identity, and if you're like most hams, you may change call signs once or twice before settling on the one you want to keep. Sometimes, your call sign starts taking over your off-the-air identity; you may become something like Ward NØAX, using your call sign in place of a last name. (I have to think really hard to remember the last names of some of my ham friends!)

Hams rarely use the term *handle* to refer to actual names; it's fallen out of favor in recent years. Similarly, they use the term *call letters* only to refer to broadcast-station licenses that have no numbers in them. Picky? Perhaps, but hams are proud of their hard-earned call signs.

Chapter 7 provides full coverage of call signs. In this section, I give you a brief overview.

Call-sign prefixes and suffixes

Each call sign is unique. Many call signs contain NØ or AX, for example, but only one call sign is NØAX. Each letter and number in a call sign is pronounced individually and not as a word — "N zero A X," for example, not "No-axe."

Hams use the Ø symbol to represent the number 0, which is a tradition from commercial operating practices.

Ham radio call signs around the world are constructed of two parts:

- **Prefix:** The prefix is composed of one or two letters and one numeral from Ø to 9. (The prefix in my call sign is NØ.) It identifies the country that issued your license and may also specify where you live within that country. For U.S. call signs, the numeral indicates the *call district* of your license when it was issued. (Mine was issued in St. Louis, Missouri, which is part of the 10th, or Ø, district.)
- **Suffix:** The *suffix* of a call sign, when added to the prefix, positively identifies you. A suffix consists of one to three letters. No punctuation characters are allowed — just letters from A to Z. (The suffix in my call sign is AX.)

The ITU assigns each country a block of prefix character groups for that government to use in creating call signs for all its radio services. All U.S. licensees (not just hams) have call signs that begin with A, K, N, or W. Even broadcast stations have call signs such as KGO or WLS. Most Canadian call signs begin with VE. English call signs may begin with G, M, or 2. Germans use D (for *Deutschland*); almost all call signs that begin with J are Japanese, and so on. You can find the complete list of ham radio prefix assignments at `www.ac6v.com/prefixes.htm#PRI`.

Class and call sign

Whatever class license you have is reflected in your assigned call sign. When you get your first license, the FCC assigns you the next call sign in the heap for your license class, in much the same way that you're assigned a license plate at the department of motor vehicles. And just as you can request a specialty license plate, you can request a special vanity call sign — within the call-sign rules, of course. The higher your license class, the shorter and more distinctive your chosen call sign can be.

Chapter 5

Studying for Your License

In This Chapter

- Breaking down the test
- Finding study resources
- Getting help from a mentor

You've decided to take the plunge and get your ham radio license. Congratulations! Although you can't just run down to the store, buy your gear, and fire it up, becoming licensed isn't a terribly difficult process. A lot of resources are available to prepare you for the ham radio exam. This chapter gives you some pointers on how best to prepare so that you enjoy studying and do well at test time.

If you buddy up with a study partner, studying is much easier. Having a partner helps you both stick with it. Each of you will find different things to be easy and difficult, so you can work together to get through the sticky spots. Best of all, you can celebrate passing the test together.

Demystifying the Test

To do the best job of studying, you need to know just what the test consists of and how it's designed. The tests for all license classes are multiple-choice; you won't find any essay questions. Some questions refer to a simple diagram. No oral questions of any kind are on the test; no one asks you to recite the standard phonetic alphabet or sing a song about Ohm's Law.

The test for each license class is called an *element.* The written exams for Technician, General, and Amateur Extra licenses (see Chapter 4) are Elements 2, 3, and 4, respectively. (Element 1 was the Morse code test, which has been dropped.)

During your studies, you'll encounter questions from the *question pool,* which is the complete set of actual questions used on the test, available to the public. The exam that you'll take is made up of a selection of questions from that pool.

The test covers four basic areas:

- **Rules & Regulations:** Important rules of the road that you have to know to operate legally
- **Operating:** Basic procedures and conventions that hams follow on the air
- **Basic Electronics:** Elementary concepts about radio waves and electronics, with some basic math involved
- **RF Safety:** Questions about how to operate and install transmitters and antennas safely

The exam must include a certain number of questions from each area; questions are selected randomly from those areas. The Technician and General tests have 35 questions; the Amateur Extra test has 50. If you answer three-quarters of the questions correctly, you pass.

Because the exam questions are public, you'll experience a strong temptation to memorize the questions and answers. Don't! Take the time to understand as much of the material as you can. After you do get your license, you'll find that studying pays off when you start operating.

Finding Resources for Study

If you're ready to start studying, what do you study? Fortunately for you, the aspiring ham, numerous study references are available to fit every taste and capability. Common study aids include classes, books, software, videos, and online help.

Before purchasing any study materials, be aware that the test questions and regulations change once every four years for each class of license. The latest changes in the Technician class questions, for example, took effect July 1, 2010, so the next set of questions will be released July 1, 2014. Be sure that any study materials you purchase include the latest updates. For the dates of the current question pools, see `www.arrl.org/question-pools` or `www.ncvec.org`.

Licensing classes

If you learn better with a group of other students, you'll find classes beneficial. You can find classes in several ways:

- **Asking at your radio club:** You can take classes sponsored by the club. If you don't see the class you want, contact the club by e-mail, and ask about classes. To find a club in your area, turn to Chapter 3.
- **Looking for upcoming exams to be held in your area:** The American Radio Relay League (ARRL) has a search engine devoted to upcoming exams at `www.arrl.org/exam_sessions/search`, as does the W5YI test-coordinator website (`www.w5yi.org`).

  Get in touch with the exam's contact liaison, and ask about licensing tests. Because tests are often given at the end of class sessions, contact liaisons are frequently class instructors themselves.

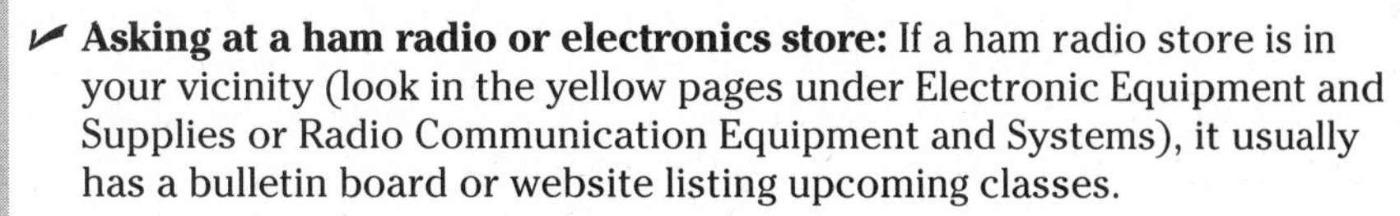

- **Asking at a ham radio or electronics store:** If a ham radio store is in your vicinity (look in the yellow pages under Electronic Equipment and Supplies or Radio Communication Equipment and Systems), it usually has a bulletin board or website listing upcoming classes.

  Businesses that sell electronics supplies to individuals, such as RadioShack, may also know about classes. In a pinch, you can do a web search for *ham radio class* or *radio licensing class* (or close variations) and your town or region.

Other options for finding classes include local disaster-preparedness organizations such as CERT (Community Emergency Response Teams, sponsored by the Federal Emergency Management Agency [FEMA]); schools and colleges, which often provide space for clubs and classes; and public-safety agencies such as police and fire departments. By asking around, you can usually turn up a reference to someone who's involved with ham radio licensing.

Occasionally, you see classes advertised that take you from interested party to successful exam-taker in a single weekend. The Technician exam is simple enough that a focused, concerted effort over a couple of days can cram enough material into your brain for you to pass the test. The good part about these sessions is that by committing a single weekend, you can walk out the door on Sunday night, having passed your exam, and find your new call sign in the Federal Communications Commission (FCC) database the following week. For busy folks or those who are in a hurry, this time savings is a tremendous incentive.

Remember when you crammed for a final exam overnight, and the minute after you took the exam, you forgot everything that was on it? The same phenomenon applies to the licensing test. A lot of information that you memorize in a short period can fade quickly. In two days, you can't really absorb the material well enough to understand it. The licensing exam isn't like high-school geometry; you'll use everything you learn in your studies later in real life. If you have time to take a weekly course, taking it is the better option.

Books, CDs, and websites

You have a variety of options in this category, including the following:

- **Guidebooks and CDs:** The best-known guidebook for licensing studies is ARRL's *Ham Radio License Manual.* Aimed at the person studying for a Technician exam, it goes well beyond presenting just the questions from the question pool; it teaches the why and how of the material. The CD bundled with the book contains practice software. A great companion to the manual is *ARRL's Tech Q&A.* Both books are available at www.arrl.org and numerous retail outlets.

  Gordon West (WB6NOA) has also written a series of licensing guides and audio courses for all three license classes. These guides focus tightly on the question pool in question-and-answer format and are geared to students who want to pass the test quickly, so a lot of the background present in the ARRL books is omitted. West's books and CDs are available at www.w5yi.org and various retail outlets.

- **Websites:** Here are a few good choices:
  - HamTestOnline (www.hamradiolicenseexam.com/index.html) offers online tutoring and training material that you can access from a web browser.
  - Ham University (www.hamuniversity.com) offers both license exam preparation and Morse code training. An ever-changing selection of ham radio study apps for the iOS and Android operating systems is available as well.
  - YouTube (www.youtube.com) hosts many ham radio tutorial videos. Search for *ham radio technician class* or *ham radio technician study guide,* and dozens of videos are yours for the clicking.

    After you get your license, YouTube videos on specific subjects are very useful as well.

Figure 5-1 shows some of these resources. For more information, visit www.artscipub.com.

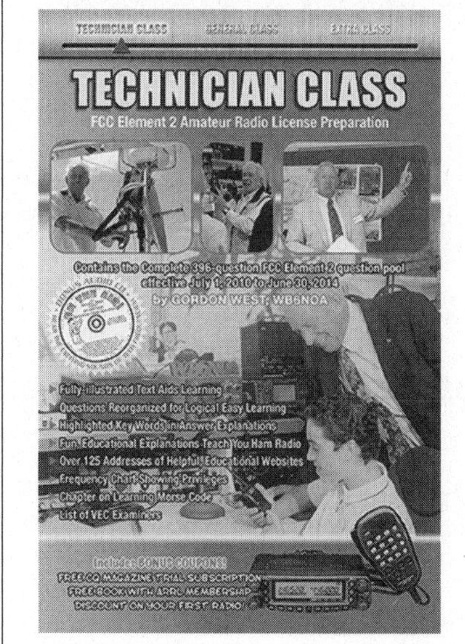

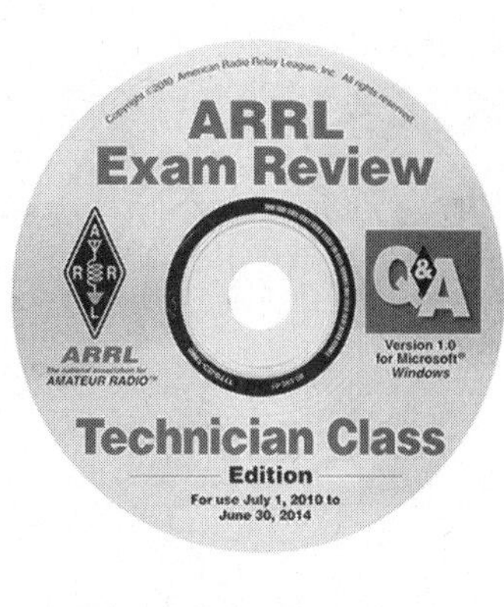

HamTestOnline™
The software that knows you™

Figure 5-1: You can use any of these resources to study for the licensing exam.

Courtesy American Radio Relay League, artscipub.com, and Master Publishing, Inc.

Online practice exams

Online practice exams can be particularly useful. When tutoring students, I urge them to practice the online exams repeatedly; because the online exams use the actual questions, they're almost like the real thing. Practicing with them reduces your nervousness and gets you used to the actual format.

The sites score your exams and let you know which of the study areas need more work. When you can pass the online exams by a comfortable margin every time, you'll do well in the actual session.

You can find online exams listed at `www.arrl.org/exam-practice`.

How do you know when it's time to stop studying and take the actual test? Take the practice exams until you consistently score 80 percent or higher. Also, make sure that you're practicing with a random selection of questions; you shouldn't see the same questions each time. Passing the practice exams with a little safety margin will give you the confidence to sign up for your test session.

Locating a Mentor

Studying for your license may take you on a journey into unfamiliar territory. You can easily get stuck at some point — maybe on a technical concept or on a rule that isn't easy to understand.

As in many similar situations, the best way to solve a problem is to call on a mentor — a more experienced person who can help you over the rough spots. In ham radio, mentors are called *Elmers.* (I discuss my own mentoring experience in the sidebar at the end of this chapter.)

Rick Lindquist (WW1ME) traces the origin of the term *Elmer* to the March 1971 issue of *QST* magazine; the term appeared in a "How's DX" column by Rod Newkirk (W9BRD). *Elmer* didn't refer to anyone specific — just to friendly, experienced hams who help new hams get their licenses and then get on the air. Nearly every ham has an Elmer at some point.

A lot of potential mentors are out there in Ham Radio Land, but you won't get far by placing a personal ad. You can find mentors in the following places:

- **Ham radio licensing classes:** Often sponsored by local ham radio clubs, classes are well worth the nominal fee (if any) you pay, if only for the personal instruction you get and the ability to have your questions answered.

- **Radio clubs:** Radio clubs can help you find classes or may even host them. Clubs welcome visitors and often have an introduction session during meetings. This session gives you an opportunity to say something like this: "Hi. My name is so-and-so. I don't have a license yet, but I'm studying and might need some help." Chances are that you'll get several offers of assistance and referrals to local experts or classes. (Find out more about clubs in Chapter 3.)
- **Online:** Although the best way to get assistance is in person, several popular ham radio websites have forums for asking questions. The eHam.net site, for example, has a licensing forum (`www.eham.net/ehamforum/smf/index.php`), and so does QRZ.com (`http://forums.qrz.com/forumdisplay.php?80-Amateur-Radio-Resources`).
- **In your community:** Many of today's hams find mentors by looking around their own neighborhoods. A ham with a tower and antenna may live near you, or you might see a car with a ham radio license plate. If you get the opportunity, introduce yourself, and explain that you're studying for a license. The person you're talking to probably also needed a mentor way back when and can give you a hand or help you find one.

After you get your license, you're in an excellent position to help other newcomers, because you know exactly how you felt at the start of the journey. Even if you're just one step ahead of the person who's asking the questions, you can be a mentor. Some hams enjoy mentoring so much that they devote much of their ham radio time to the job. You won't find a higher compliment in ham radio than being called "my Elmer."

My mentor experience

When I started in ham radio, my mentor was Bill (then WNØDYV, now KJ7PC), a fellow high-school student who had been licensed for a year or so. I wasn't having any trouble with the electronics, but I sure needed a hand with the Morse code and some of the rules. I spent every Thursday over at Bill's house, practicing Morse code *(pounding brass)* and learning to recognize my personal-nemesis characters: D, U, G, and W. Without his help, my path to getting licensed would have been considerably longer. Thanks, Bill!

Since getting my license, I've required the assistance of several other mentors as I entered new aspects of ham radio. If you can count on the help of a mentor, your road to a license will be much smoother.

Chapter 6
Taking the Test

In This Chapter

- Searching for a test session
- Registering for a test
- Arriving at the Big Day

After your diligent studies, you find yourself easily passing the online tests by a comfortable margin. Now you're ready to — drumroll, please! — take the test.

If you're part of a class or study group, the test may be part of the planned program. In this case, you're all set; just show up on time. Skip to the last section of this chapter.

If you're studying on your own, however, read on. This chapter tells you where and when you can take the test.

Finding a Test Session

Fortunately for you, finding a schedule of test sessions in your area is pretty easy. The Federal Communications Commissions (FCC) provides a list of organizations that serve as Volunteer Examiner Coordinators (VECs) at `http://wireless.fcc.gov/services/index.htm?job=licensing_5&id=amateur`. (For details on the connection between VECs and licensing exams, see Chapter 4.)

If a VEC in the list is close to you, start by contacting that organization. Many VECs have websites, and all of them have e-mail contacts. Visit the website or send an e-mail that says something like this: "My name is . . . and I want to take the Technician (or General or Amateur Extra) class license exam. Please send me a list of examination sites and dates. I live in . . ."

If you don't see a nearby VEC, or if no exams are scheduled at times or places that are suitable for you, you can find an exam conducted by one of these national VEC organizations:

- **ARRL VEC exams:** The American Radio Relay League (ARRL) has a VEC that operates nationwide. Search for exams based on your zip code at `www.arrl.org/exam_sessions/search`.
- **W5YI VEC exams:** Like ARRL, W5YI (founded by Fred Maia [W5YI]) has a nationwide VEC. You can find a list of certified examiners to contact at `www.w5yi-vec.org/exam_locations_ama.php`.
- **W4VEC VEC exams:** W4VEC (the call sign of the Volunteer Examiners Club of America) covers the Midwest and South. For a list of dates and locations, visit `www.w4vec.com/ar.html`.

If you still can't find an exam that's convenient for you, your final option is to write or e-mail VECs at the addresses listed on the aforementioned FCC website and ask for help. The mission of VECs is to help prospective amateurs get licensed. No matter where you live, these organizations can put you in touch with Volunteer Examiners (VEs; see Chapter 4) so that you can take your test.

Signing Up for a Test

After you find a test session, contact the test session hosts or sponsors to let them know that you'll be attending the session and what test elements you want to take.

Checking in ahead of time isn't just good manners, but also can alert you to time or location changes.

Public exams

Most exam sessions are open to the public and are held at schools, churches, and other public meeting places. Nearly all sessions are open to walk-ins — that is, you can just show up unannounced, pay your test fee, and take the test — but some require an appointment or reservation. Checking before you show up is always a good idea.

Call or e-mail the session's contact person to confirm the date, get directions, and tell him or her what tests you need to take.

Exams at events

Exam sessions are common at public events such as hamfests and conventions. (See Chapter 3 for information on these events.) These sessions can attract dozens of examinees and often fill up quickly. Some exams are given more than once throughout the day, so you can take more than one test or can spend time enjoying the event.

Under FCC rules, you're not required to pay an attendance fee for the event if you're going just to take a license exam, but you may encounter a special entry fee. Don't be afraid to call ahead and ask, or check the event's website.

If you attend an event-sponsored exam, it's a good idea to get to the site early to register. Multiple sessions may be offered, and the tests for different elements may be given only at specific times.

Private exams

Small test sessions may be held in private residences, especially in rural areas and small towns. When I was a VE living in a small town, most of the new hams in my community passed their exams while sitting at my kitchen table.

If you'll be taking your test at a private residence, call ahead to ensure that there'll be room for you at the session and that the VE can prepare to administer the exam you want to take.

You'll be a guest in someone's home, so act accordingly.

Getting to Test Day

In the so-called Good Old Days, the higher-class license exam sessions were conducted in federal office buildings by FCC employees. I vividly recall standing in line with dozens of other hams, waiting for my shot at a new license. Some of us drove for hours to reach the FCC office or test location, nervously reviewing the material or listening to Morse code tapes between swallows of coffee. Inside, a steely-eyed examiner watched as we scratched out the answers.

These days, exams are certainly more conveniently offered and the examiners are friendlier, but you'll still have some nervous anticipation as the day arrives.

The best way to do well, of course, is to be prepared — for all aspects of the exams, not just the questions. The more you know, the less you have to worry about.

For some advice on getting ready for the big day, see the nearby sidebar "Taming the test tiger."

What to bring with you

Be sure to bring these items with you to the test session, whether you're licensed or not:

- Two forms of identification, including at least one photo ID, such as a driver's license or employer's identity card
- Your Social Security number
- A couple of pencils
- A calculator
- (Optional) Scratch paper (but it must be completely blank)

If you already have a license and are taking an exam to upgrade to a higher class, you also need to bring the following:

- Your current original license and a photocopy
- Any original Certificate of Successful Completion of Examination (CSCE) you have and a photocopy

  Note: This certificate is your record of having passed an exam for one or more of the license elements. If you've just passed the Technician exam (Element 2), you have to wait for the FCC to grant you a call sign before you get on the air. For any other license changes, the CSCE allows you to operate immediately with your new privileges.
- (Optional) Your FCC Federal Registration Number (FRN) as a substitute for your Social Security number

  In Chapter 7, I discuss why you might want to do this.

You aren't permitted to use any kind of online device or computer during the exam unless you have a disability (and you must first coordinate the use of supporting devices with the test administrators).

Taming the test tiger

Follow these surefire pointers to turn that tiger of a test into a pussycat by keeping your thinker in top shape:

Do:

- Wear a couple of layers of clothing to make yourself comfortable, whatever the room temperature.
- Visit the restroom before the session starts.
- Follow the directions for completing the identification part of your answer sheet, even though you may want to start the test right away.
- Study a question that seems to be really difficult; then move on to the next one. When you come back to the question later, it may be crystal-clear.
- Completely erase the wrong answer or indicate clearly that you made a change if you change any answers.
- Double-check your answers before handing in your test to make sure that you marked the answers you wanted.

Don't:

- Take the test when you're hungry, sleepy, or thirsty.
- Drink extra coffee or tea.
- Change an answer unless you're quite sure. (Generally, your first choice is your best choice.)
- Leave a question unanswered. Guess if you have absolutely no idea about the answer; you have a one-in-four chance of getting it right. If you leave the answer sheet blank, you have no chance of getting the right answer.
- Rush. Remember to breathe; take a minute to stretch, roll your head, or flex your arms and legs. The test isn't a race. Take your time.

What to expect

Each test session involves three basic steps:

1. **Register for your exam.**

   When you arrive at the test session, sign in with your name, address, and (if you already have one) call sign. The test administrators review any identification and documents you have. Finally, pay your test fee. (As of early 2013, the largest VECs charge a $15 test fee.)

2. **Take the test.**

   When you start depends on how many people have signed up ahead of you and how many types of tests are being given. In a small session, you may start the test immediately; in a larger session, you may have to wait a while until your turn comes.

As I discuss in Chapter 5, the exams are multiple-choice tests. You receive a pamphlet containing the test questions and an answer sheet for recording your choices.

Each test takes 15 to 45 minutes. The session may be organized so that everyone starts and stops together, or the testing may be continuous. The VEs will explain the process for your session.

3. **Complete your paperwork (which I talk about in Chapter 7).**

What to do after the test

When you're done with your exam, follow the administrator's instructions for turning in your paper, sit back, and try to exhale! Depending on the size of the session, you may have to wait several minutes for the administrator to grade your paper. At lease three VEs verify the grades on all exams. Passing requires a score of 75 percent or better. (That's 26 questions on the Technician and General exams and 37 on the Extra.)

In all probability, because you studied hard and seriously, you'll get a big smile and a thumbs-up from the test graders. Way to go! You can finally, truly relax and move on to the next stage.

If you didn't pass this time, don't be disheartened. Many sessions allow you to take a different version of the test, if you want. Even if you don't take the second-chance exam again right away, at least you know the ropes of a test session now, and you'll be more relaxed next time.

Don't let a failure stop you. Many hams had to make more than one attempt to pass a test, but they're on the air today.

Chapter 7

Obtaining Your License and Call Sign

In This Chapter

- Filling out your paperwork
- Watching for your call sign
- Choosing your own call sign
- Keeping your license valid

After you pass your exam (see Chapter 6), only a small matter of paperwork separates you from your new license. The exam-session volunteers help you complete everything correctly and even send your paperwork to the Federal Communications Commission (FCC).

You still need to understand what you're filling out, though; that's what I cover in this chapter. Fill your paperwork out correctly, and you won't delay the process of getting your call sign.

Completing Your Licensing Paperwork

After you successfully complete the exam, you need to fill out two forms:

- **Certificate of Successful Completion of Examination (CSCE):** Figure 7-1 shows the ARRL-VEC CSCE. (As I discuss in Chapter 6, this certificate is issued by the American Radio Relay League's Volunteer Examiner Coordinator.) The VEC and FCC use the CSCE as a check against the test-session records. Your copy of the completed form documents your test results and shows credit for the exam you passed at any other test session before you receive your license or upgrade from the FCC.

VEC: American Radio Relay League/VEC
CERTIFICATE of SUCCESSFUL COMPLETION of EXAMINATION

Test Site (city/state): ______________________ Test Date: ______________

CREDIT for ELEMENTS PASSED
You have passed the telegraphy and/or written element(s) indicated at right. You will be given credit for the appropriate examination element(s), for up to 365 days from the date shown at the top of this certificate, if you wish to upgrade your license class again while a newly-upgraded license application is pending with the FCC.

LICENSE UPGRADE NOTICE
If you also hold a valid FCC-issued Amateur radio license grant, this Certificate validates temporary operation with the operating privileges of your new operator class (see Sction 97.9[b] of the FCC's Rules) until you are granted the license for your new operator class, or for a period of 365 days from the test date stated above on this certificate, whichever comes first. **Note:** If you hold a current FCC-granted (codeless) Technician class operator license, and if this certificate indicates Element 1 credit, this certificate indefinitely permits you HF operating privileges as specified in Section 97.301(e) of the FCC rules. This document must be kept indefinitely with your Technician class operator license in order to use these privileges.

LICENSE STATUS INQUIRIES
You can find out if a new license or upgrade has been "granted" by the FCC. For on-line inquiries see the FCC Web at **http://www.fcc.gov/wtb/uls** ("License Search" tab), or see the ARRL Web at **http://www.arrl.org/fcc/fcclook.php3**; or by calling FCC toll free at 888-225-5322; or by calling the ARRL at 1-860-594-0300 during business hours. Allow 15 days from the test date before calling.

THIS CERTIFICATE IS NOT A LICENSE, PERMIT, OR ANY OTHER KIND OF OPERATING AUTHORITY IN AND OF ITSELF. THE ELEMENT CREDITS AND/OR OPERATING PRIVILEGES THAT MAY BE INDICATED IN THE LICENSE UPGRADE NOTICE ARE VALID FOR 365 DAYS FROM THE TEST DATE. THE HOLDER NAMED HEREON MUST ALSO HAVE BEEN GRANTED AN AMATEUR RADIO LICENSE ISSUED BY THE FCC TO OPERATE ON THE AIR.

The applicant named herein has presented the following valid exam element credit(s) in order to qualify for the license earned category indicated below:
Circle the **bold** text from one or more of these examples:
–for pre 3/21/87 Technicians circle **3/21/87 Tech-EL 1+3**;
–for pre 2/14/91 Technicans circle **2/14/91 Tech-El 1**;
–for lifetime Novice code credit circle **Novice-El 1**;
–for a valid or expired-less-than-5-years FCC Radiotelegraph license/permit circle **FCC Telegraph-EL 1**;

NOTE TO VE TEAM: **COMPLETELY** CROSS OUT ALL BOXES BELOW THAT DO NOT APPLY TO THIS CANDIDATE.

EXAM ELEMENTS EARNED
passed 5 wpm code element 1
passed written element 2
passed written element 3
passed written element 4

NEW LICENSE CLASS EARNED
TECHNICIAN
TECHNICIAN w/HF
GENERAL
EXTRA

Candidate's signature ______________________
Candidate's name ______________ Call sign ________ (if none, write none)
Address ______________________
City ______________ State ________ ZIP ________

VE #1 ______ signature ______ call sign
VE #2 ______ signature ______ call sign
VE #3 ______ signature ______ call sign

Candidate's copy=white•ARRL/VEC's copy=pink•VE Team's copy=yellow

Figure 7-1: The ARRL VEC CSCE.

Keep your copy of the CSCE until the FCC sends you a new license or records the change in its database. You'll probably want to hang on to it as a record of your achievement.

- **NCVEC Form 605:** NCVEC Form 605 allows the FCC to process your new license. Whenever you get a new license, upgrade to a higher class, renew your license, change your name or address, or pick a new call sign, you use this form. You can also submit name, address, or call-sign changes directly to the FCC by mail or online.

You're required to maintain your current mailing address on file with the FCC. Any time your mailing address changes, keep the FCC database up to date. Mail sent to the address in the FCC database should get to you in ten days or fewer.

The Volunteer Examiners (VEs) who run the test session send your completed CSCE and NCVEC Form 605 to the certifying VEC organization. Asking your examiners about the average wait before the FCC updates your information in its database is a good idea. On average, the wait is about seven to ten business days.

Finding Your Call Sign

If you took the test to upgrade to a higher-class license, you already have a call sign, but you'll need to change it temporarily (see "Identifying with your new privileges," later in this chapter).

If you're a first-time licensee, you can begin watching the FCC database for your new call sign to appear after you complete the test session and your paperwork is sent to the VEC. The next two sections walk you through the search process.

When your call sign appears in the FCC database with your name beside it, you can get on the air even if you don't have the paper license.

Searching the ULS database

The FCC has an online licensee information database called the Universal Licensing System (ULS), shown in Figure 7-2. Although you can search other databases (see the next section), the one maintained by the FCC is the one that really counts.

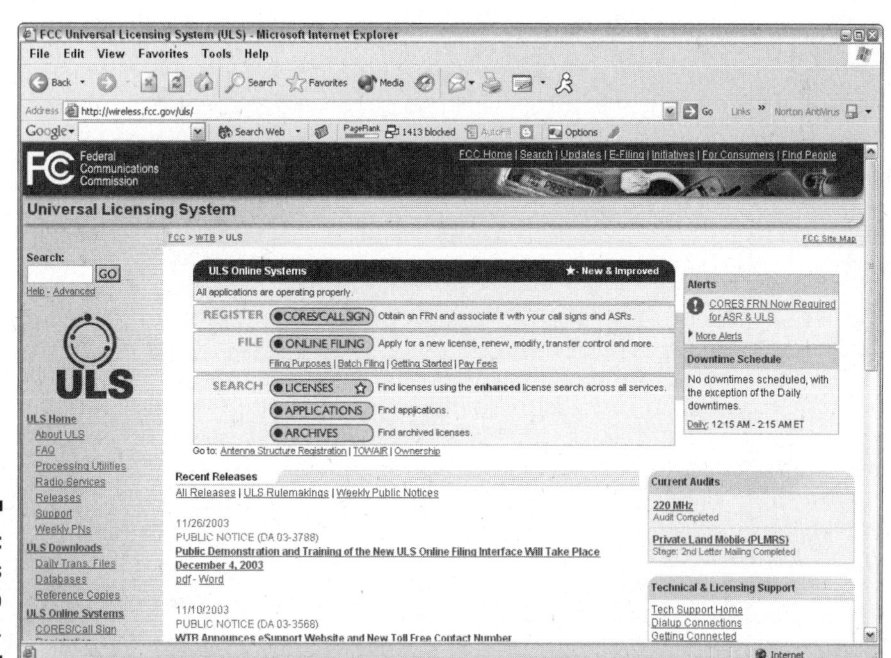

Figure 7-2: The FCC's ULS web page.

Each licensee has a Federal Registration Number (FRN) that serves as identification within the FCC. I outline the process of registering for your own FRN in "Registering with the FCC Online," later in this chapter.

Follow these steps to find your call sign online:

1. **Log on to the ULS system at** `http://wireless.fcc.gov/uls` **(refer to Figure 7-2).**
2. **In the Search section, click the Licenses button.**

   The License Search page loads (see Figure 7-3).
3. **Click the Amateur link in the Service Specific Search column in the middle of the page.**

   You may have to scroll to see it.
4. **In the Licensee section, enter your last name and zip code in the appropriate boxes.**
5. **Click the Search button in the bottom-right corner.**

   It may take a few seconds for your request to be processed and the search result to appear. Figure 7-4 shows the result of a search for *Silver* and *98070.*

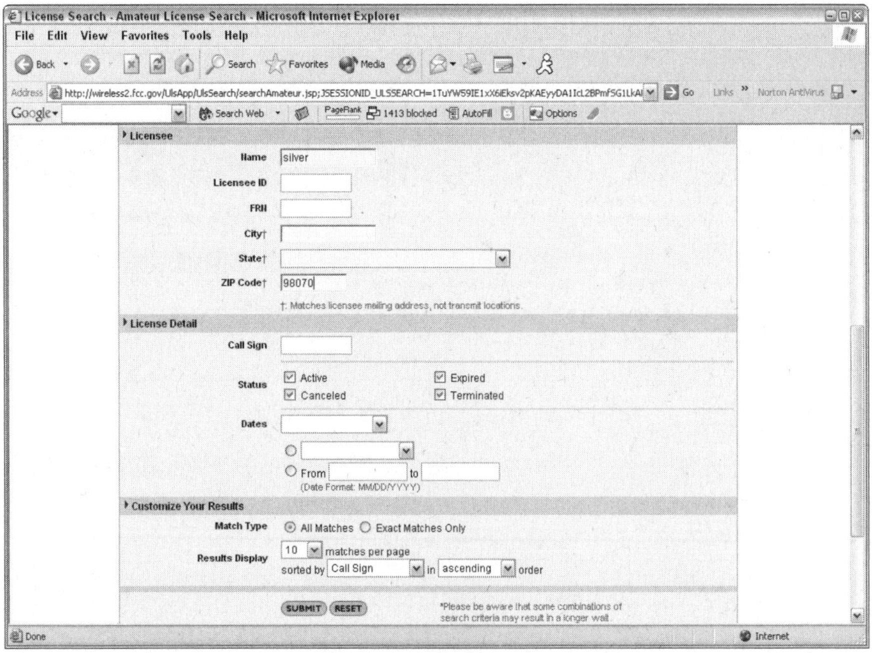

Figure 7-3: The FCC's ULS License Search page.

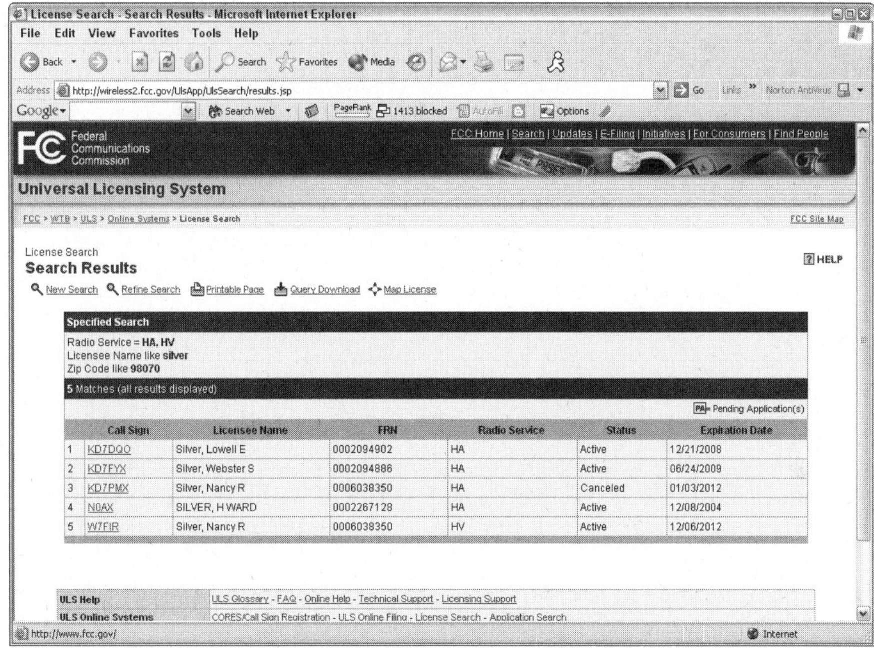

Figure 7-4: The search result page lists my information.

6. **Browse the result page.**

   If the results take up more than one page, click the Query Download link to have the entire batch of results compiled into a single text file.

Feel free to browse through the database. By doing a little creative investigating in the License Search page, you can discover some interesting things about the ham population in your area. It's fun to see how many hams have the same last name as you, for example, or how many live in your zip code.

Keep in mind that anyone else can do the same sort of search. When you submit your application, you identify yourself as a licensed ham and indicate where your *fixed station* (a station that doesn't move) is located. This information is made available to the public because you obtained a license to use the public airwaves; therefore, the public is entitled to know who's licensed, and with what call sign and FRN. Your personal information, however, is password-protected. I discuss this topic in detail in "Registering with the FCC Online," later in this chapter.

Searching other call-sign databases

You can also browse a couple of other websites that access the FCC database:

- **QRZ.com** (`www.qrz.com`)**:** QRZ.com is the best-known ham radio call-sign-lookup website. You need to register to gain access to call sign information. After you do, type the call sign in the search box in the top-left corner of the home page and then click Search.

  If you don't know the call sign, you can search by name, address, county, or grid square by making a choice from the menu below the search box. (I explain *grid square* in Chapter 11.)

  You can also search for non-U.S. hams if data from their countries is available online.

- **ARRL** (`www.arrl.org/fcc/search`)**:** Type your last name and zip code, and click the Search button for results.

Identifying with your new privileges

If you're upgrading an existing license, which means that you already have a call sign, you can go home after passing the test and use your new privileges right away. You just have to add a temporary suffix to your call sign to let everyone know that you're qualified to use those privileges.

Here are the suffixes you need to add to your call sign:

- **Upgrade to General:** Add /AG to your call sign on Morse code or digital modes, and "slash AG" or "temporary AG" on voice.
- **Upgrade to Amateur Extra:** Add /AE to your call sign on Morse code or digital modes, and "slash AE" or "temporary AE" on voice.

When your new license comes in the mail or your new license class is recorded in the FCC database, you can drop the temporary suffix.

Registering with the FCC Online

The FCC has done a lot of work to make ordinary license transactions easier by creating the online ULS system, which allows you to process renewals, address changes, and other simple services. To use this system, however, you need to register in the Commission Registration System (CORES), whether or not the FCC has already granted your license.

What if you don't find your call sign?

Patience is difficult while you're waiting, but be sure to wait at least one full calendar week before getting worried. If two weeks pass, you can take action. Here's what you can do:

- Contact the leader of the exam session, and ask whether the VEC accepted your paperwork and sent it to the FCC. Some problem may have caused a delay in the FCC's acceptance of the session results.
- If the paperwork went through okay, and it's been more than ten business days (or longer than the usual wait for the VEC that coordinated your session), ask the session leader to inquire about your paperwork. The VEC can trace all applications to the FCC.

Problems are rare, however. In more than ten years of being a session leader, I never had any paperwork lost or delayed without reason.

Registering in CORES

Follow these steps to register with CORES as an individual amateur licensee:

1. **Go to** `http://wireless.fcc.gov/uls`.

   The ULS web page appears (refer to Figure 7-2, earlier in this chapter).

2. **Click the Register button.**

   An options page appears.

3. **Select one of the following options and then click the Continue button:**

   - *Register Now:* Select this option if you're using the ULS for the first time.
   - *Update Registration Information:* If any of your information has changed since the last time you used the system, select this option.
   - *Update Call Sign/ASR Information:* Select this option if you need to add any call signs to or remove any call signs from the list that you associated with your FRN. (Hams rarely use this option.)

4. **In the next page, select the An Individual option and your contact address; then click the Continue button.**

5. **Enter your name and address in the page that appears.**

   Any field marked with an asterisk is a required field. Your telephone number, fax number, and e-mail address are optional.

   Everything you enter on this page except your Social Security number, phone number, and e-mail address will be available for public inspection.

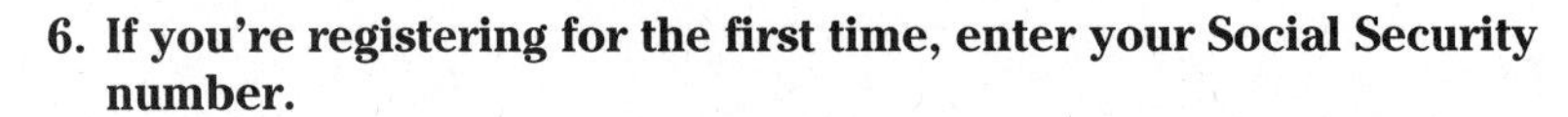

6. **If you're registering for the first time, enter your Social Security number.**

   You're required to provide this information (or give a reason why you can't). Enter the number without any spaces, hyphens, or periods.

   Ignore any prompts or windows asking for a Sub-Group Identification Number (SGIN), which is used by managers of large communications services that have many call signs. You don't need a SGIN.

7. **At the bottom of the window, enter a password of 6 to 15 characters (or have the system pick one for you); then enter it again in the Re-Enter Password box.**

   Don't use your call sign (or any part of it) as your password, because an unauthorized person would try it immediately.

8. **In the Hint box, enter a password reminder.**

   If you ever forget your password and want the FCC to tell you what it is, this hint verifies that you're you — not someone else. You can enter any word or words that fit in the box.

9. **Click the Submit button.**

   The ULS system processes your entries and displays a page that lists any errors you made, such as omitting a required item or entering the wrong type of information in a particular field.

10. **Correct any errors, and click the Submit button again.**

    The system displays a form containing your licensee information, password, and password hint.

    Print this information and keep a copy in a safe place in case you ever forget your password.

You're now registered with the FCC. The next step is associating your call sign with your FRN, so leave this page open in your web browser and proceed to the next section.

Associating your call sign with your ID

Follow these steps to associate your call sign with your FRN:

1. **Click the FCC Universal Licensing System link at the bottom of your licensee information page (see Step 10 in the preceding section).**
2. **Click the Call Sign/ASR Registration link.**

   A window loads, with your FRN already entered.

3. **Enter your password in the Password box, and click the Continue button.**
4. **In the next page, click the Enter Call Signs link.**

   The Enter Call Signs window loads.
5. **Enter your call sign in the first space provided, and click the Submit button.**

Now both you and your call sign are registered with the FCC. Numerous services are available to you for free, such as renewing your license or making an address change.

Picking Your Own Call Sign

You can pick your own call sign (within certain limits, of course). If you're the sort of person who likes having a license plate that says IMABOZO or UTURKEY, you'll enjoy creating a so-called *vanity call sign.*

Short call signs and ones that seem to spell words are highly sought. Many hams enjoy having calls made up of their initials. Whatever your preference, you'll likely find a vanity call sign that works for you.

Come up with at least two candidate call signs before you file an application (see "Applying for a vanity call sign," later in this chapter). That way, if your first choice is unavailable, you still have options.

Searching for available call signs

You can find available call signs by using the FCC's ULS search function (see "Searching the ULS database," earlier in this chapter), but that system can be quite cumbersome because it's designed to return information on only one call sign at a time. The following websites offer better and more flexible call-sign-search capabilities:

- **N4MC's Vanity HQ (**`www.vanityhq.com`**):** By using N4MC's website, you can quickly determine which call signs are available for assignment.
- **WM7D call-sign database (**`www.wm7d.net/fcc_uls`**):** This site offers a good search function that allows wildcard characters, which speeds your search for that perfect call sign.

Finding call signs available to you

Depending on your license class (see Chapter 4), you can select any available call sign in the groups listed in Table 7-1. (***Note:*** No new Novice and Advanced licenses are being issued, as I discuss in Chapter 4.)

Call signs are referred to as *2-by-3 (2x3)* or *1-by-2 (1x2)*, meaning the number of letters in the prefix (first) and the suffix (second). KDØPES is a 2 (KD) by 3 (PES) call, for example, and NØAX is a 1 (N) by 2 (AX) call. The FCC assigns certain types of call signs to the various license classes, with the higher-class licensees having access to the shorter and presumably more attractive calls.

Table 7-1 explains the structure of call signs, broken down by license class.

Table 7-1 Call Signs Available by License Class

License Class	*Types of Available Call Signs*
Technician and General	2x3, with a prefix of KA–KG, KI–KK, KM–KO, or KR–KZ and a suffix of any three letters
	1x3, with a prefix of K, N, or W and a suffix of any three letters
Amateur Extra	2x3, with a prefix of KA–KG, KI–KK, KM–KO, or KR–KZ and a suffix of any three letters
	2x1, with a prefix beginning with A, K, N, or W and a suffix of any letter
	1x2, with a prefix of K, N, or W and a suffix of any two letters
	1x3, with a prefix of K, N, or W and a suffix of any three letters
Novice	2x3, with a prefix of KA–KG, KI–KK, KM–KO, or KR–KZ and a suffix of any three letters
Advanced	2x2, with a prefix of K or W and a suffix of any three letters

Note: This table doesn't cover call signs in Alaska, Hawaii, or the various U.S. possessions in the Caribbean and Pacific. Special rules apply to those locations.

Occasionally, you hear a call sign consisting of one letter, one numeral, and one number. Call signs of this type — called *1-by-1 (1x1)* call signs — are granted on a temporary basis to U.S. hams for expeditions, conventions, public events, and other noteworthy activities. The special call-sign program is administered by several VEC organizations for the FCC. For more information, visit `www.arrl.org/special-event-call-signs`.

Ham radio license plates

You can also acquire a license plate with your call sign. The process is easy, and many states even have a special type of vanity plate just for hams. Contact your local department of motor vehicles and ask! For additional information, see www.arrl.org/amateur-license-plate-information.

One wrinkle for hams with call signs that contain the slashed-zero (Ø) character: In most states, you have to request the slashed zero specifically. Talk to the clerk who handles your form, and show him or her examples of ham call signs with that character. Pictures of ham license plates are very effective in explaining what you're asking for.

Applying for a vanity call sign

When you've narrowed down your list of candidate vanity call signs, follow these steps to file your application:

1. **Go to** www.arrl.org/vanity-call-signs.
2. **List one or more call signs that you like.**

   All the call signs must be unassigned and available (which is why you need to search the vanity call websites first; see "Searching for available call signs," earlier in this chapter).
3. **Fill out the rest of the application.**
4. **Pay the $15 fee (as of December 2012) by credit card or check.**

Maintaining Your License

The FCC supports a set of common filing tasks at http://wireless.fcc.gov/services. This page is the place to go if you want to do any of the following things:

- Renew your license.
- Change any of the information associated with the license, such as your name or address.
- Replace your physical license.
- Check on an application.
- Apply for a vanity call sign (but it's a lot easier to use the ARRL VEC).

Check out this page when you first earn your license so that you'll be familiar with it when you need to take care of any licensing business later.

Part III

Hamming It Up

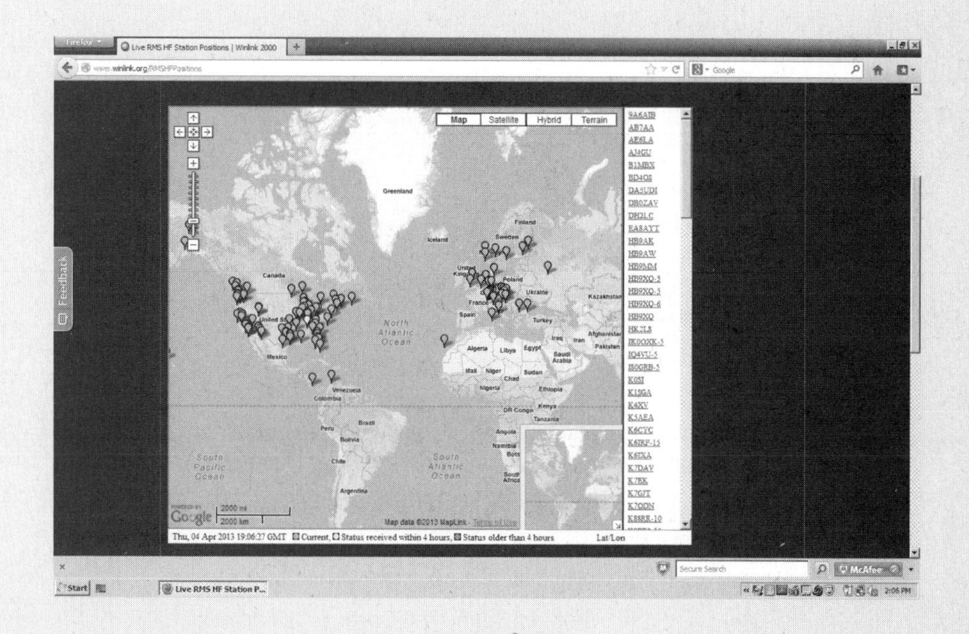

On the Web Extras page at www.dummies.com/extras/hamradio, discover a free downloadable program that helps you predict when you'll be able to communicate with stations in a specific part of the world, near or far.

In this part . . .

- Find out how, where, and when to listen, listen, listen — a ham's most important skill.
- Understand how to interpret what you hear on the air and how to make a contact yourself.
- Get acquainted with the ham bands and how their characteristics change through the day.
- Step through making contacts by voice, Morse code, and even sending e-mail.
- Discover several interesting operating specialties.
- Find out about the different types of ham emergency communications and how you can join in.
- On the Web Extras page at `www.dummies.com/extras/hamradio`, discover a free downloadable program that helps you predict when you'll be able to communicate with stations in a specific part of the world, near or far.

Chapter 8
Making Contact

In This Chapter

- Listening on different bands
- Finding the right signal
- Interpreting a contact
- Making a call

When you have your ham radio *(rig)* set up and a license *(ticket)* clearing you for takeoff, you're ready to make your first connection. If this thought makes your palms a little sweaty, don't worry; all hams start out feeling just that way, and they survive. You will, too.

In this chapter, I show you how to make a contact. I also cover the basics of on-the-air manners and the simple methods that make contacts flow smoothly. With a little preparation, you'll feel comfortable and confident, ready to get on the air and join the fun.

Listen, Listen, Listen!

The most important part of successfully putting a contact, also known as a *QSO,* in your logbook is listening — or, in the case of the digital modes, watching what the computer displays. (*QSO* is a *Q-signal* — one of several ham radio abbreviations explained in the sidebar "Q&A with Q-signals," later in the chapter.) In fact, your ears (and eyes) are the most powerful parts of your station. The ham bands are like a 24-hour-a-day party, with people coming and going all the time. Just as you do when you walk into any other big party, you need to size up the room by doing two things for a while before jumping in:

- *Tuning the band* (receiving on different frequencies to assess activity)
- *Monitoring* (listening to or watching an ongoing contact or conversation)

By doing so, you discover who's out there and what they're doing, what the radio conditions are like, and what the best way for you to make contact is.

Listening on different bands

You can listen on the following bands:

- HF *(high frequency)* bands cover 3 MHz to 30 MHz and are usually thought of as the *shortwave* bands.
- VHF *(very high frequency)* bands cover 30 MHz to 300 MHz.
- UHF *(ultra high frequency)* bands cover 300 MHz to 3 GHz.
- *Microwaves* are considered to start at about 1 GHz.

The shortwave or HF bands have a different flavor from the VHF bands. On the HF bands, you can find stations on any frequency that offer a clear spot for a contact. Up on the VHF bands, most contacts take place by means of repeaters on specific frequencies or on channels spaced regularly by a few kHz. How are you supposed to figure out where the other hams hang out?

As a Technician licensee, you're likely to listen on the VHF and UHF bands at first, but don't miss an opportunity to take in what's happening on the lower-frequency HF bands, which have a completely different flavor.

Repeaters are radios that listen on one frequency and retransmit what they hear on another frequency. Repeaters usually are located at high spots such as hilltops or tall towers so that they can pick up weak signals well, and they have powerful transmitters so that their signals can be heard a long distance away. Repeaters allow weak portable and mobile stations to communicate over a wide area. Repeaters are most useful on VHF and higher-frequency bands.

On both HF and VHF, hams engage in specific activities and tend to congregate on or near specific frequencies. Digital fans who use the popular PSK31 mode, for example, usually hang out near 14.070 MHz. No rule says that they *must* operate on that frequency, but they gather there routinely anyway. That kind of consistency provides a convenient way for you to meet others who have similar interests and equipment. To continue my party metaphor, it's like when a fellow partygoer tells you, "A group is usually talking about jazz at that table in the corner." Whenever groups tend to congregate at particular frequencies, those frequencies are known as *calling frequencies.* When a frequency becomes known as a spot on the band where you can find other hams using similar modes or operating styles, it's a calling frequency.

Understanding sub-bands and band plans

In the United States, regulations specify where each type of signal may be transmitted in a given band. These segments of the band are called *sub-bands.* Figure 8-1 shows the sub-bands for the 80 meter band. The American Radio Relay League (ARRL) offers a handy chart of U.S. sub-bands at `www.arrl.org/files/file/hambands_color.pdf`.

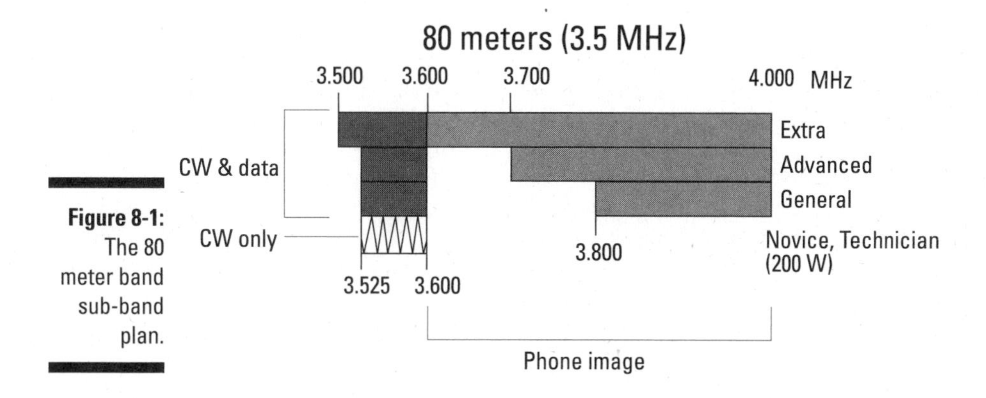

Figure 8-1: The 80 meter band sub-band plan.

Outside the United States, regulations are much less restrictive. You'll probably hear Canadian and overseas hams having voice contacts in a part of the 40 meter band where American hams don't have phone-transmitting privileges. (*Phone* is an abbreviation for *radiotelephone,* which includes all voice modes of transmission.) How unfair! Because of the number of American hams, the Federal Communications Commission (FCC) long ago decided that to maintain order, it was necessary to segregate the wide-bandwidth phone signals from narrow-bandwidth code and digital or data signals. That's just the way it is. So close and yet so far!

Beyond this segregation of the amateur bands, hams have collectively organized themselves to organize the different operating styles on each band. Not all amateur users can coexist on the same frequency, so agreements about where different types of operations occur are necessary. These agreements are called *band plans.* Band plans are based on FCC regulations but go beyond them to recognize popular calling frequencies and segments of a band where you usually find certain operating styles or modes.

A band plan isn't a regulation and should be considered to apply only during normal conditions. When a lot of activity is going on — such as during emergency operations, a contest, or even a big expedition to a rare country — don't expect the band plans to be followed. Be flexible and work around the activity, or jump in and participate.

A complete list of all band plans is beyond the scope of this book, but a good source of up-to-date U.S. band plans is `www.arrl.org/band-plan`.

Be aware that band plans may be different outside the United States. Europe and Japan, for example, have substantial differences on certain bands.

Tuning In a Signal

Tuning in a signal consists of using the main tuning control on the radio to change the radio's operating frequency. If you're using a traditional radio, the control is a rotating knob or dial. On a PC-based radio that uses software to receive the signal or on some digital modes, the control may be a mouse-operated slider. Either way, on the HF bands, this control tunes the radio smoothly across a range of frequencies. In the repeater sub-bands at VHF and UHF, the tuning control jumps from channel to channel. Usually, when you know what frequency you want, you have a way to enter that frequency directly.

The frequency control usually changes the frequency of a *variable-frequency oscillator* (VFO), which actually controls the radio's frequency. Even though software-type radios don't have VFO circuits, hams still refer to the "VFO control" as a matter of convenience. The radio's operating frequency is presented on some kind of numeric display. A received signal's frequency is the radio's displayed frequency when the signal is tuned in properly.

The correct method of tuning in a signal depends on the type of signal it is. On signals that you receive or copy manually — such as Morse code (CW), single sideband (SSB), or FM — you use your ears. For digital or data signals — such as radioteletype (RTTY), packet, or PSK31 — you use the characters shown on the computer screen get your receiver set just right. There may be *tuning indicators* to help you zero in on a digital signal. Whether you tune in a signal from above or below its frequency doesn't matter, although you may develop a preference for one or the other.

Tuning in a signal begins with selecting the type of signal you want to receive, such as PSK31, RTTY, SSB, or maybe CW. Set your radio to receive this type of signal as described in the operating manual (you *do* have your operating manual handy, right?) or the software help file. The next few sections give you an idea of how to tune in a few types of signals. Here you go!

Listening on SSB

Single sideband (SSB) is the most popular mode of voice transmission on the HF bands. (FM is mainly used above 50 MHz.) The mode got its name from a key difference from the older mode, AM, which is used by AM broadcast stations and was the original voice mode that hams used. Whereas an AM transmitter outputs two identical copies of the voice information, called *sidebands,* a SSB signal outputs only one. This signal is much more efficient and saves precious radio-spectrum space. (AM still has a dedicated following of hams who appreciate the mode's characteristic fidelity and equipment, however.)

Most voice signals on HF are SSB, so you have to choose between USB *upper sideband (USB)* and *lower sideband (LSB).* The actual SSB signals extend in a narrow band above (USB) or below (LSB) the carrier frequency displayed on the radio (see Figure 8-2).

How do you choose? By long tradition stemming from the design of the early sideband rigs, on the HF bands above 9 MHz, the convention is that voice operation takes place using USB. Below 9 MHz, you find everyone on LSB except (by FCC decree) on 60 meters.

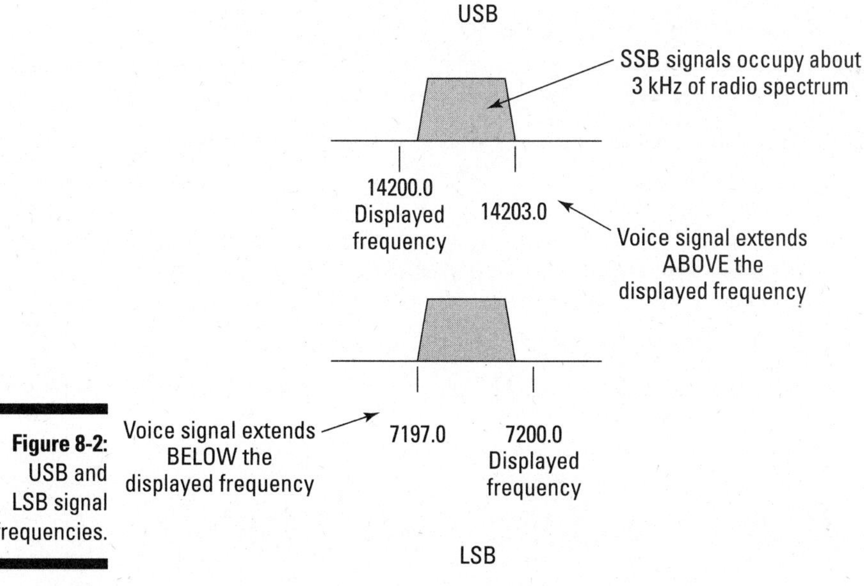

Figure 8-2: USB and LSB signal frequencies.

Because hams must keep all signals within the allocated bands, you need to remember where your signal is actually transmitted. Most voice signals occupy about 3 kHz of bandwidth. If the radio is set to USB, your signal appears on the air starting at the displayed frequency up to 3 kHz higher. Similarly, on LSB, the signal appears up to 3 kHz below the displayed frequency. When you're operating close to the band edges, make sure that your signal stays in the allocated band. On 20 meters, for example, the highest frequency allowed for ham signals is 14.350 MHz. When transmitting a USB signal, you may tune your radio no higher than 14.350 MHz — 3 kHz = 14.347 MHz to keep all your signal inside the band and stay legal.

If you tune across an AM signal that has both upper and lower sidebands, you can tune in the signal with either USB or LSB. The whistling noise that gets lower and lower in frequency as you tune in the voice is the AM carrier signal, centered precisely between the two voice sidebands. When the pitch of the carrier tone goes so low that you can't hear it anymore, you can listen to either sideband by switching between USB and LSB.

To tune in an SSB signal, follow these steps:

1. **Set your rig to receive SSB signals.**

   You may have to choose LSB or USB.

2. **Select the widest SSB filter.**

   To select filters, you use a Wide/Narrow control or buttons labeled with filter widths. (Check the operating manual for precise instructions.)

3. **Adjust the tuning dial until you hear the SSB frequency.**

   As you approach an SSB signal's frequency, you hear either high-pitched crackling (like quacking) or low-pitched rumbling. You can tell from the rhythm that you're listening to a human voice, but the words are unintelligible. What you're hearing are the high- and low-frequency parts of the operator's voice.

4. **Continue to tune until the voice sounds natural.**

   If the voice sounds too bassy, your transmitted signal sounds too trebley to the receiving operator, and vice versa.

Listening to digital or data signals

Many types of digital signals exist, and all of them sound a little different on the air, from buzzing (PSK31) to two-tone chatter (RTTY) to a sound like a rambling calliope (MFSK). Each type of signal requires a different technique to tune in, so in this section, I focus on an easy-to-use mode that you're likely to try right away: PSK31.

First, you need to be running some digital-mode decoding software. By far the most popular program in 2013 is FLDIGI, by W1HKJ. You can download it for free at `www.w1hkj.com/Fldigi.html` and start using it immediately, as described in the software's help file. If you start using the digital modes regularly, you'll want to use an audio interface gadget between the radio and your PC. But to give the mode a try, here's a way to listen to a digital mode signal: Just use the computer's microphone and turn up the radio's volume.

When you have the software installed, follow these steps to tune in PSK31:

1. **Set your rig to USB, and tune to one of the PSK31 calling frequencies.**

   You might try a frequency such as 14.070 MHz during the day or 7.035 MHz at night.

2. **If several filters are available, select one that's suitable for voice, such as the standard 2.4 kHz filter.**

   (For details on selecting filters, see "Listening on SSB," earlier in this chapter.)

   If PSK31 signals are present, you'll hear whistling or buzzing. If you hear a hiss or static, try a different band.

3. **Run FLDIGI, and set it to receive PSK31.**

   (Refer to FLDIGI's instructions to get started.)

4. **Turn on the waterfall display.**

   This display shows the signal as a yellow stripe against a blue-and-black background (see Figure 8-3).

5. **Adjust the receive volume until the background is mostly covered with blue speckles.**

   Assuming that signals are present, you'll see them as yellow stripes slowly making their way down the page.

6. **Click the strongest signal.**

   The red PSK31 channel markers straddle the signal, and if it's a PSK31 signal, the decoded text is displayed in the yellow window.

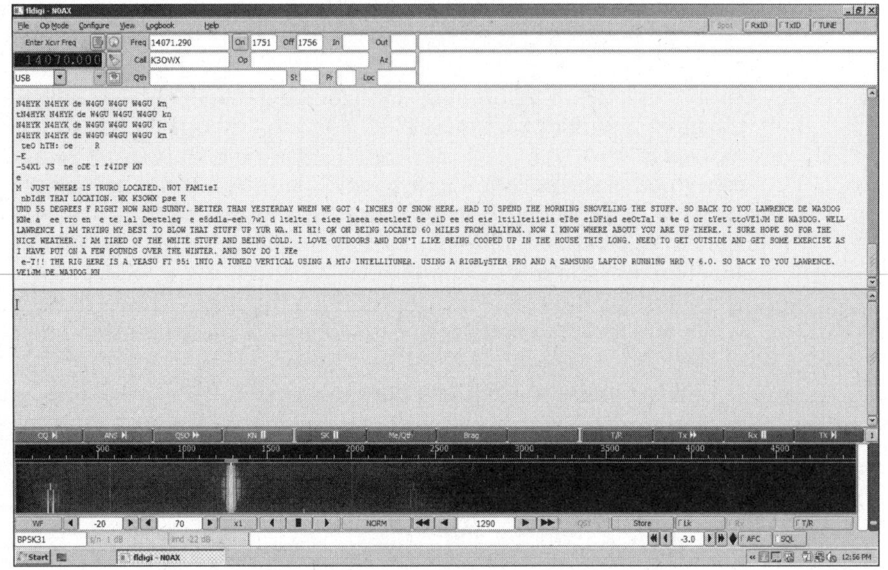

Figure 8-3: The FLDIGI display with received text and a waterfall-style signal display.

That's really all there is to it: Run the software, connect the audio, and click a signal to start receiving. Practice this technique to receive other PSK31 signals. You may be surprised by what you can hear, because PSK31 is a very efficient mode.

When you get good at PSK31, try tuning around the RTTY or MFSK calling frequencies, and see whether you can tune in a few of them.

Identifying bands and modes by ear or eye

With so many types of digital modes to choose among and with more being invented all the time, how can you tell what you're hearing or seeing on the bands? W1HKJ, the provider of the popular FLDIGI software, recorded signals using most of the popular digital modes and posted them at `www.w1hkj.com/FldigiHelp-3.12/Modes/index.htm` to help you practice identifying them by ear or eye.

If you like to listen outside the ham bands, try `www.kb9ukd.com/digital` for samples of commercial and military modes.

Listening to FM channels

Frequency modulation (FM) is the most popular mode of transmission on the VHF and UHF bands. FM signals transmit a voice as frequency variations. Because hams adapted surplus FM radios from businesses and public-safety agencies to the ham bands, operation has been organized as *channels* on specific frequencies. As a result, tuning on most FM rigs consists of selecting different channels or moving between specific frequencies instead of making a continuous frequency adjustment.

To tune in an FM signal, follow these steps:

1. **(Optional) Set your rig to operate on FM.**

   Most VHF/UHF radios use only FM, so your radio may not have a control for selecting the mode.

2. **Set the squelch control so that you hear noise.**

   This procedure is called *opening the squelch.*

3. **Reset the squelch so that it just stops the noise.**

   This step enables you to hear weak signals without having to listen to continuous noise. For very weak signals, you may have to reopen the squelch to receive them.

4. **Using a band plan or a repeater directory as a guide, select a frequency.**

   If you're using an FM-only rig such as a handheld or mobile unit, you can enter the frequency via a keypad, rotate a small tuning control that changes frequency (as on HF), or select a memory channel. If hams are active on that channel, you hear the operator's voice. Depending on the change in frequency with each step, you may have to tune back and forth to find the frequency where the voices sound best. If you're mis-tuned *(off frequency)*, the voices are muffled or distorted.

   If you have a multiple-band radio that uses a main tuning knob or slider, you can tune in the signal by using your ear (tuning for the most natural-sounding voice with the least distortion) or, if available, by watching a tuning indicator called a *discriminator,* sometimes labeled DISC. The dis-criminator shows whether you're above or below the FM signal's center frequency. When the signal is centered, you're tuned just right.

Listening to Morse code

Morse code signals are often referred to as *CW,* which stands for *continuous wave.* Early radio signals died out quickly because they were generated by sparks. Soon, however, operators discovered how to make steady signals, or continuous waves, by turning the signals on and off with a telegraph key. Thus, *Morse code* and *CW* became synonymous.

To tune in a Morse code signal, follow these steps:

1. **Set the rig to receive Morse code by selecting the CW mode and tuning to a frequency somewhere in the bottom 20 kHz to 30 kHz of an HF band.**
2. **If your rig has more than one filter, set it to use a wide filter.**

   A wide filter allows you to find and tune in stations, whereas the narrower ones block out unwanted nearby signals. You select filters with a Wide/Narrow control or with buttons labeled with filter widths.
3. **Adjust the tuning control until you hear a Morse code signal.**

   The pitch changes as you change the receiver's frequency. Tune until the pitch is comfortable to your ear.

   A low tone (300–600 Hz) is most restful to the ear, but a higher tone (500–1200 Hz) often sounds crisper. Most radios are designed so that when you tune in a signal with a tone or pitch around 600 Hz, the transmitted signal is heard by the other station at a similar pitch. If you prefer to listen to a note more than 100 Hz higher or lower, check your rig's operating manual to find out how you can adjust the radio to accommodate your preferred pitch.
4. **When you tune in the signal at your preferred pitch, select a narrower filter (if one is available) to reduce noise and interference.**

   If the frequency isn't crowded or noisy, you can stay with a wider filter.

Listening on HF

Most of the traditional shortwave bands between 1.8 MHz and 30 MHz are broadly organized into two segments. In the United States, Morse code (CW) and data signals occupy the lower segment, and voice signals occupy the higher segment. Within each of these segments, the lower frequencies are where you tend to find the long-distance *(DX)* contacts, special-event stations, and contest operating. Casual conversations *(ragchews)* and scheduled on-the-air meetings *(nets)* generally take place on the higher frequencies within each band.

Organizing activity on HF bands

Table 8-1 provides some general guidelines on where you can find different types of activity. Depending on which activity holds your interest, start at one edge of the listed frequency ranges and start tuning as described in "Tuning In a Signal," earlier in this chapter.

While tuning, use the widest filters your radio has for the mode (CW, SSB, or FM) that you select. That way, you won't miss a station if you tune quickly, and finding the right frequency when you discover a contact is easier. After you tune in a contact, you can tighten your filters to narrower bandwidths, limiting what you hear to just the one contact.

Table 8-1 Activity Map for the HF Bands

Band	*CW, RTTY, and Data Modes*	*Voice and Image Modes*
160 meters (1.8–2.0 MHz)	1.800–1.860 MHz (no fixed top limit)	1.840–2.000 MHz
80 meters (3.5–4.0 MHz)	3.500–3.600 MHz	3.600–4.000 MHz
60 meters (5.3-5.4 MHz	Permitted, but the signal has to be centered in the channel.	5330.5, 5346.5, 5357.0, 5371.5, and 5403.5 MHz (voice, CW, RTTY, and data only)
40 meters (7.0–7.3 MHz)	7.000–7.125 MHz	7.125–7.300 MHz
30 meters (10.1–10.15 MHz)	10.100–10.125 MHz CW 10.125–10.150 MHz RTTY and data	Not permitted
20 meters (14.0–14.35 MHz)	14.000–14.150 MHz	14.150–14.350 MHz
17 meters (18.068–18.168 MHz)	18.068–18.100 MHz (no fixed top limit)	18.110–18.168 MHz
15 meters (21.0–21.45 MHz)	21.000–21.200 MHz	21.200–21.450 MHz
12 meters (24.89–24.99 MHz	24.890–24.930 MHz (no fixed top limit)	24.930–24.990 MHz
10 meters (28.0–29.7 MHz)	28.000–28.300 MHz	28.300–29.7 MHz (most activity below 28.600 MHz)

If every voice that you hear sounds scrambled, your rig is probably set to receive the wrong sideband. Change sidebands, and try again.

Because hams share the 60 meter band with government stations, there are special rules for operating on this band. Read the rules for 60 meter operation before getting on the air.

Adjusting for time of day

Because the ionosphere strongly affects signals on the HF bands as they travel from point A to point B, the time of day makes a big difference. On the lower bands, the lower layers of the ionosphere absorb signals by day but disappear at night, allowing signals to reflect off the higher layers for long distances. Conversely, the higher bands require the Sun's illumination for the layers to reflect HF signals back to Earth, supporting long-distance *hops* or *skips.* (With the exception of sporadic effects, the ionosphere is much less a factor on the VHF and UHF bands at 50 MHz and above.)

Table 8-2 shows general guidelines on what you might hear on different HF bands at different times of day.

Table 8-2 Day/Night HF Band Use

HF Band	*Day*	*Night*
160, 80, and 60 meters (1.8, 3.5, and 5 MHz)	Local and regional out to 100–200 miles.	Local to long distance, with DX best near sunset or sunrise at one end or both ends of the contact.
40 and 30 meters (7 and 10 MHz)	Local and regional out to 300–400 miles.	Short-range (20 or 30 miles) and medium distances (150 miles) to worldwide.
20 and 17 meters (14 and 18 MHz)	Regional to long distance; bands open at or near sunrise and close at night.	20 meters: Often open to the west at night and may be open 24 hours a day. 17 meters: Follows the same pattern but opens a little later and closes a little earlier.
15, 12, and 10 meters (21, 24, and 28 MHz)	Primarily long distance (1,000 miles or more); bands open to the east after sunrise and to the west in the afternoon.	15 meters: A good daytime band, especially to the Caribbean and South America, closing right after sunset. 12 meters and 10 meters: Usually have short openings in the morning and afternoon (unless there are lots of sunspots). 10 meters: Often used for local communications 24 hours a day.

Finding and using beacons

After you tune the bands, you may still not know for sure whether the band is *open* (meaning that signals can travel beyond line of sight), and in what direction. Propagation software is available to make predictions that help with those decisions, but those predictions are only predictions. To help you determine whether a band is actually open, beacon transmitters are set up around the world. A *beacon* sends a message continuously on a published frequency. Amateurs who receive that beacon's signal know that the band is open to its location.

The biggest network of beacons in the world is run by the Northern California DX Foundation (NCDXF) and the International Amateur Radio Union (IARU). The network consists of 18 beacons around the world, as shown in the following figure. These beacons transmit on the 20 through 10 meter bands in a round-robin sequence. They also vary their transmitting power from 100 watts to 100 milliwatts so that hams receiving the beacon signal can judge the quality of propagation. A complete description of this useful network is available at `www.ncdxf.org/beacons`.

Other amateurs have set up beacons, too. You can find lists of frequencies for these amateur beacons on various websites. A good reference for all beacons on HF and 6 meters is the excellent list at `www.keele.ac.uk/depts/por/28.htm`. Amateur VHF beacons are listed on several websites; just enter *amateur VHF beacon* in a web search engine to locate several beacon listings.

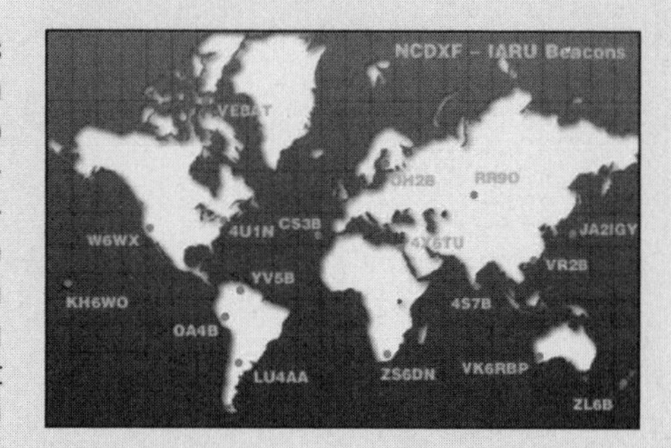

Listening on VHF and UHF

Most contacts on the VHF and UHF bands are made with repeaters (see Chapter 2). Repeaters are most useful for local and regional communication, allowing individual hams to use low-power handheld or mobile radios to make contacts over that same wide area. For this scheme to work, the repeater input and output frequencies are fixed and well known, so the bands are organized into sets of channels. (Except for some at the upper end of the 10 meter band, repeaters aren't used on the HF bands because of the need to receive and transmit simultaneously — difficult within single HF bands.)

Most VHF and UHF voice contacts use the FM mode of voice transmission because of its excellent noise suppression, making for comfortable listening. The drawback is that FM doesn't have the range of CW or SSB transmissions. Contacts made directly between hams via FM are referred to as *simplex,* and those made via a repeater are *duplex.*

Mapping contacts online

Because most hams have a computer online right in their shacks, they can report what they're hearing and who they're contacting. Several sites use this information to plot contacts between stations on a map. When the bands are open, contact after contact pops up. Sometimes, a band may be closed at your location, but by watching the online map, you can see propagation gradually moving in your direction.

The best-known set of online contact maps is at www.dxmaps.com/spots/map.php, which is run by Gabriel Sampol (EA6VQ), from the Balearic Islands, off the coast of Spain. You can watch maps of contacts on most amateur bands, send messages to other stations, check solar and ionospheric data, and do much more.

If you'd rather run a mapping application directly on your computer, you can use the program ViewProp (http://zl2ham.wikispaces.com), by Rick Kiessig (ZL2HAM), from New Zealand.

As the online mapping tools become better and better, look for them to be combined with real-time information from the ham bands.

Repeater and simplex FM channels are generally separated by 15 or 20 kHz. You can view a complete band plan for the 2 meter and 70 cm bands at www.arrl.org/band-plan. I cover repeater operation in more detail in Chapter 9.

Repeaters enable you to use low-power and mobile radios to communicate over a large distance. Many hams use repeaters as kinds of intercoms to keep in touch with friends and family members as they go about their daily business. These contacts generally are much less formal than those on HF, and you're likely to hear contacts among the same groups of hams every day. Repeaters are where you find local hams and find out about local events.

Finding contacts via repeater

To make contact via a repeater, you may have to enable tone access on your radio. *Tone access* adds one of several standard low-frequency tones to your speech audio to let the repeater know that your signal is intended for it and isn't interference. If you don't transmit the required tone, the repeater doesn't retransmit your signal, and you can't be heard. (The radio's operating manual can tell you how to select and activate the tones.)

Not all repeater channels have an active repeater. To find repeaters in your area or while traveling, check a repeater directory or website (see Figure 8-4). Some of these directories are nationwide, such as the *ARRL*

Repeater Directory, available as a book or on the TravelPlus CD-ROM (`www.arrl.org/shop`); others focus on specific regions, such as the New England Repeater Directory at `www.nerepeaters.com`. RF Finder (`www.rfinder.net`) is a smartphone app that accesses a worldwide directory on a subscription basis. The directories list the frequencies and locations of repeaters so you can tell which ones may be available in your area. Repeater directories also list the required access tones and the other operating information and features for individual repeaters.

Figure 8-4: Repeater directories are available online, as books, or as CD-ROMs.

Courtesy American Radio Relay League

Listening to repeater contacts

To listen to repeater contacts, follow these steps:

1. **Use a repeater directory or website to find a repeater in your area.**

   For details, see the preceding section.

2. **Determine the repeater's input and output frequencies.**

3. **Set up your radio to listen on the repeater's output frequency.**

   You can also listen to stations transmitting to the repeater — an act called *listening on the input.*

4. **Tune your radio as you do for FM signals.**

   If you need help tuning your radio, turn to the section "Tuning In a Signal," earlier in this chapter.

Some VHF/UHF radios have an "auto repeater" feature that can tell when you are tuned to the repeater channels and automatically set up the radio to transmit and receive on the different frequencies required.

Making weak-signal contacts

For direct ham-to-ham contacts on VHF and UHF over distances at which FM results in noisy, unpleasant contacts, use the more efficient CW and SSB modes. This method is called *weak-signal* communication on VHF and UHF because you can make contacts with much lower signal levels than you can by using FM. The lowest segments of the VHF and UHF bands are set aside for weak-signal operation.

Weak-signal operations are conducted in much the same way as SSB and CW operations on HF, with contacts taking place on semirandom frequencies centered on calling frequencies. For details on tuning in contacts on the weak-signal segments of VHF and UHF bands, refer to "Tuning In a Signal," earlier in this chapter.

Deciphering QSOs

As you tune across the bands, dozens of contacts (called *QSOs)* may be going on. That number may sound like a bewildering variety, but you'll find that most QSOs are one of three types:

- Casual conversation *(ragchews)*
- Nets
- Contesting *(DXing)*

I discuss all three types in the following sections.

Chewing the rag

Chewing the rag is probably the oldest type of activity in ham radio. I have no idea where the expression came from (except possibly from *chewing the fat*), but if you like to chat, you're a *ragchewer* and are following in the footsteps of the mythical master ragchewer The Old Sock himself. Ragchewing is an excellent way to build your operating skills, perhaps leading to an award such as the A-1 Operator's Club Award, shown in Figure 8-5.

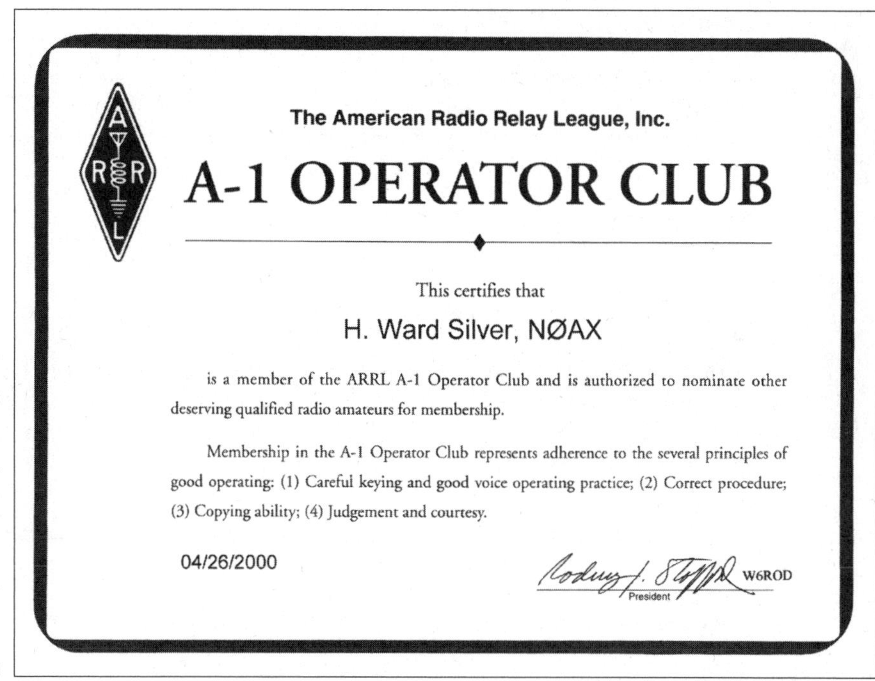

The American Radio Relay League, Inc.

A-1 OPERATOR CLUB

This certifies that

H. Ward Silver, NØAX

is a member of the ARRL A-1 Operator Club and is authorized to nominate other deserving qualified radio amateurs for membership.

Membership in the A-1 Operator Club represents adherence to the several principles of good operating: (1) Careful keying and good voice operating practice; (2) Correct procedure; (3) Copying ability; (4) Judgement and courtesy.

04/26/2000

President W6ROD

Figure 8-5: The ARRL offers several operating awards, including this one.

Keep these things in mind while you're chewing the rag:

- **Start with your basic information: your call sign, signal report, operator name, and station location.** Ragchews may be conducted between hams in the same town or a world apart.

Most digital mode software encourages you to enter information about yourself and your station that's stored as messages — called *brag macros* — that are sent by pressing a single key.

- **After you exchange basic information, you may wander off in any direction.** Hams talk about family members, other hobbies, work, propagation, technical topics, operating, you name it. Just about anything may be discussed.

  In general, hams avoid talking about politics or religious topics, and they don't use profanity. Those restrictions still leave a lot to talk about, however, and hams seem to cover most of it.

- **Wrap up the contact when you run out of things to talk about, when conditions change, or when maintaining contact gets difficult.** Exchange call signs once more, and tune away.

Meeting on nets

Nets (short for *networks)* meet at a regular time and on a consistent frequency. Each net has a theme — perhaps message handling *(traffic)*, emergency-communications training, maritime or mobile service, or specific topics such as antique radios or technical Q&A.

You can find nets for a specific topic or frequency online. The ARRL Net Directory, for example, is at `www.arrl.org/arrl-net-directory`.

Follow these tips when accessing a net:

- **Tune in to the listed net frequency, listen for a request to check in with the net control station (NCS), and list your business.** The NCS orchestrates all exchanges of information and formally terminates the net when business is concluded. (The net may meet until all business is taken care of or for a fixed amount of time.)
- **If you're a visitor, find out when you can check in.** Nets often have specific times when visitor stations can participate.

I discuss nets and net operation in more detail in Chapter 10.

Similar to nets are *roundtables,* in which a group of hams on one frequency share information informally. Each ham transmits in turn, and everyone gets a chance to transmit in sequence. Some roundtables have moderators.

Contesting and DXing

Many hams like to participate in radio contests, such as competitions in which they exchange call signs and short messages as quickly as possible.

Similar to short-term contests is *chasing DX* or *DXing* — pursuing contacts with distant stations.

While you're contesting or DXing, follow these rules:

- **Keep contest contacts short.** In a contest, the object generally is to make the largest number of contacts, so dilly-dallying around isn't desirable.
- **Pass along just the minimum amount of information, called the *exchange*.** Then sign off in search of more contacts.

  While you're DXing, keep contacts with rare stations short if many stations are calling; exchange just call signs and signal reports. Keeping contacts short allows other hams to make those sought-after contacts too.

  DX contacts tend to be short because the distances are great and maintaining contact is difficult. Ragchews with DX stations are encouraged only if conditions support good signals in both directions. Try to judge conditions, and tailor your contact appropriately.

- **If you encounter stations making contest QSOs, listen until you figure out what information is being exchanged before participating.** By far the most common types of information exchanged are signal reports, locations (often expressed as a numbered zone or section defined by the contest sponsor), and serial numbers.

  Serial numbers are assigned for each contact made in the contest. If you're making your fifth QSO in a contest, for example, your serial number is 5.

  If the rate of making contacts is relatively relaxed, contesters are happy to explain what information they need. Usually, you can find the complete rules for contests in online contest calendars such as `www.hornucopia.com/contestcal` and `www.arrl.org/contest-calendar`, which includes *QST* magazine's Contest Corral, or on the sponsor's website. If contestants are making contacts lickety-split, you may want to wait, or figure out what they need on your own and then make a quick contact.

- **Remember that not everyone speaks English.** Most hams who don't speak English still know enough words of English to communicate a name, location, and signal report. Otherwise, an international set of Q-signals (see the nearby sidebar) allows you to exchange a lot of useful information with people who speak a different language.

  DXing is a great way to exercise that rusty high-school Spanish or German.

I cover contests and DXing in more detail in Chapter 11.

Q&A with Q-signals

Q-signals began in the early days of radio as a set of standard abbreviations to save time and to allow radio operators who didn't speak a common language to communicate effectively. Today, amateurs use Q-signals as shorthand to speed communication. The definitions have drifted a little over the past century of radio, but the use of these signals is ubiquitous. Table 8-3 lists many common Q-signals.

During contacts, Q-signals often take the form of questions. "QTH?" means "What is your location?" for example, and the reply "QTH New York" means "My location is New York." Sometimes, Q-signals are just an abbreviation, such as QSO, which you already know means "contact." QST is "calling all amateurs."

Table 8-3 Common Q-Signals

Q-Signal	*Meaning As a Query*	*Meaning As a Response*
QRG	What is my exact frequency?	Your exact frequency is _______ kHz.
QRL	Is the frequency busy?	The frequency is busy. Please don't interfere.
QRM	Are you being interfered with?	I'm being interfered with (or just Interference).
QRN	Are you receiving static?	I am receiving static (or just Static).
QRO	Shall I increase power?	Increase power.
QRP	Shall I decrease power?	Decrease power.
QRQ	Shall I send faster?	Send faster (__words per minute [wpm]).
QRS	Shall I send more slowly?	Send more slowly (__wpm).
QRT	Shall I stop sending?	Stop sending.
QRU	Have you anything more for me?	I have nothing more for you.
QRV	Are you ready?	I'm ready.
QRX	Do you want me to stand by?	Stand by.
QRZ	Who's calling me?	(Not used as a response)
QSB	Is my signal fading?	Your signal is fading.
QSL	Did you receive and understand my transmission?	Your transmission was received and understood.
QSP	Can you relay to ___?	I can relay to____.
QSX	Can you receive on ___ kHz?	I'm listening on ___ kHz.
QSY	Can you change to transmit on another frequency (or to ___ kHz)?	I can transmit on another frequency (or to ___ kHz).
QTC	Do you have messages for me?	I have messages for you.
QTH	What is your location?	My location is ____.

Making Your Own Contacts

Your big moment approaches! In the next two sections, I walk you through the process of initiating contacts, using my call sign and my son's — NØAX and KD7FYX, respectively — as examples. (Don't forget that Ø is the way hams write a zero.) Replace my call sign with your own.

Making contact on an HF band

Follow these steps when tuning one of the HF bands:

1. **Find someone to talk to.**

   For this example, use a voice contact. When you come across a fellow ham who's making a general "Come in, anybody" call, you've found someone who's *calling CQ* (see "Calling CQ," later in this chapter). This situation is the easiest way for you to make a contact. You'll hear something like this: "CQ CQ CQ, this is November Zero Alfa X-ray, standing by . . ."

   November, Alfa, and *X-ray* are *phonetics* that represent the letters of my call sign. Phonetics are used because many letters sound alike (think *B, E, T,* and *P*), and the words help get the exact call sign across.

   Table 8-4 lists the standard International Telecommunication Union (ITU) phonetics that hams use. You may encounter alternatives, such as *Germany* instead of *Golf.* When in doubt, I respond or call with the phonetics used by the station I want to contact.

2. **Carefully note the other ham's call sign, and respond *(come back).***

   Press the microphone button and say something like this: "November Zero Alfa X-ray, this is Kilo Delta Seven Foxtrot Yankee X-ray [repeat twice more], over."

   Give the calling station's call sign once (you don't have to repeat it; the other ham already knows it) and then give yours three times — a setup known as a *1-by-3 call.* If the calling station is strong, you can give your call twice instead of three times — a *1-by-2 call.*

   You don't need to shout. Just speak in a normal, clear voice.

3. **Listen for a response.**

   You may hear something like this: "KD7FYX [possibly in phonetics] from NØAX, thanks for the call. Your signal report is. . . ." When you do, you have a QSO in your logbook!

   If you don't get a response, turn to "Failing to make contact," later in this chapter.

Table 8-4 ITU Standard Phonetics

Letter	Phonetic	Letter	Phonetic
A	Alfa	N	November
B	Bravo	O	Oscar
C	Charlie	P	Papa
D	Delta	Q	Quebec
E	Echo	R	Romeo
F	Foxtrot	S	Sierra
G	Golf	T	Tango
H	Hotel	U	Uniform
I	India	V	Victor
J	Juliet	W	Whiskey
K	Kilo	X	X-ray
L	Lima	Y	Yankee
M	Mike	Z	Zulu

Making contact via repeater

Making contact works a little differently on a repeater. Hams use repeaters mostly as regional intercoms, so they're less likely to make general calls for random QSOs. (You never hear "CQ CQ CQ" on a repeater.) Hams turn the radio on in the car or in the shack to listen for friends or just monitor the channel.

If someone is available for a contact, that person announces his or her presence by saying something like "This is NØAX, monitoring," or making some other kind of general "I'm here" announcement. Just send out a 1-by-1 call consisting of your call followed by their call ("NØAX, this is KD7FYX") to see whether the other ham comes back to you. You also get good results by calling a station immediately after it completes a contact.

To respond on a repeater, where signals probably are quite clear, just make a 1-by-1 call ("NØAX, this is KD7FYX").

Repeater use tends to be more utility-oriented than on HF, so you may find the contacts to be a little briefer.

Making contact via Morse code or digital mode

With CW and digital modes, the process is much the same as it is for HF bands (see "Making contact on an HF band," earlier in this chapter).

To make an initial contact, follow these steps:

1. **Copy the calling station's call sign.**

   You hear (or see, if you're using software) something like this: "CQ CQ CQ DE NØAX NØAX NØAX K."

   DE is telegrapher's shorthand for *from. K* means "end of transmission, go ahead." (***Note:*** Morse code doesn't use uppercase or lowercase characters, so *de* is equivalent to *DE.*)

2. **Respond with a 1-by-2 or 1-by-3 call.**

   Send or type something like this: "NØAX DE KD7FYX KD7FYX KD7FYX K." (For details on 1-by-2 and 1-by-3 calls, see "Making contact on an HF band," earlier in this chapter.)

3. **Listen for a response.**

   The response will be something like this: "KD7FYX DE NØAX TKS FOR THE CALL MY NAME IS. . . ." (*TKS* is shorthand for *thanks,* and *UR* is shorthand for *your.*)

   If you don't get a response, see the next section.

Telegraphers and typists are a lazy lot who tend to use all sorts of abbreviations to shorten text. A table of abbreviations is available at `http://ac6v.com/morseaids.htm#CW`.

Failing to make contact

What if you try to make a contact and your call doesn't get a response? Your signal may be too weak, or the station may have strong noise or interference. In such a case, just find another station to call.

Assuming that your signal is strong enough for other stations to hear, however, several other things may have happened:

- **Other hams are calling at the same time.** You can either wait around until the station you intend to contact is free and then try again, or you can tune around for another contact opportunity. The most important thing is to keep from getting discouraged!

- **The calling station can hear you but can't make out your call.** The ham may either ask you to call again or respond to you, but he or she won't have your call sign correct. The station may say or send "Station calling, please come again" or "QRZed?" or "Who is the station calling?"

  QRZed? is the international Q-signal for "Who's calling me?" (refer to Table 8-3, earlier in this chapter). Hams often use the British pronunciation of the letter *Z,* which is *zed.* On the digital modes or Morse code, you would receive just *QRZ?*

  At this point, just repeat your call two or three times, using standard phonetics, and say "Over" when you finish.

- **The station gets your call wrong by a letter or two.** First, stand by for a few seconds to make sure that another station with a similar call sign isn't on the same frequency. (I'm often on the air at the same time as NØAXL and NØXA, for example, and we're always getting confused.) If a few seconds go by and you don't hear another station responding, respond as follows: "NØAX, this is KD7FYX [repeat twice more]. Do you have my call correct? Over."

If repeated attempts at making contacts aren't producing results, check out your equipment. The easiest way is to locate a licensed friend and have him or her make a contact with you. That way, you'll know whether your transmitter is working and your signal is understandable. If you don't have a licensed friend, run through the following checklist to make sure you're transmitting what, when, and where you think you should be:

- **Are you transmitting on the right frequency?** Press the microphone switch or press the Morse key, and watch the radio's display very carefully. The indicators for frequency and sideband should stay exactly the same. If not, you're transmitting on a different sideband or frequency from the one you think you're using.
- **Is your transmitter is producing power?** Watch the rig's power output meter to ensure that the output power varies along with your voice or keying.
- **Is the antenna connected properly?** You should be receiving most signals as moderate to strong, with an indication of 4 to 9 on the radio's S meter, which displays signal strength. If the signals are very weak, you may have an antenna or cable problem. This problem also shows up as a reading of more than 5:1 on your rig's SWR meter, which measures how well your transmitter power is getting to the antenna. (A reading of 1:1 is the best case, and values over 2:1 indicate that you may have an antenna or feed line problem.)

After your call is received correctly by the other station, proceed with the rest of the contact.

Failed contacts and errors are handled very similarly on CW and the digital modes.

Don't be bashful about correcting your call sign. After all, it's your radio name.

Breaking in

Sometimes, you can't wait for the end of a contact to call a station. Interrupting another contact is called *breaking in* (or *breaking*). The proper procedure is to wait for a pause in the contact and quickly say "Break" (or send *BK* with Morse code) followed by your call sign.

Why do you want to do this? Perhaps you have an emergency and need to make contact right away. More frequently, you tune in to a contact, and the participants are talking about a topic with which you're familiar; if you wait for the contact to end, you may not be able to contribute or help.

To break in a contact, follow these steps:

1. **Listen for a good opportunity to make your presence known.**

   When the stations switch transmitting and receiving roles, that's usually a good time to break in. You may hear something like this: "So, Sharon, back to you. AE7SD from NØAX."

2. **Quickly make a short transmission.**

   Don't be shy and wait for the other station to begin transmitting. Say "Break," followed by your call sign, just once.

3. **Wait to see whether either station heard your transmission.**

   If a station hears you, the operator may say something like this: "This is AE7SD. Who's the breaker?"

   If no one hears your transmission, start over with Step 1.

4. **Respond as though you're answering a CQ.**

   Say this: "AE7SD, this is KD7FYX [repeated twice]. Over."

5. **Depending on the circumstances, give your name and location before proceeding to explain why you broke in.**

   At that point, the stations may engage you in further conversation, and you'll be in a three-way QSO. Sometimes, however, they won't want to have a third party in the contact, in which case you should just courteously sign off and go on to the next contact.

Having a QSO

Because you listen to contacts (QSOs) on the air, you understand the general flow of the contacts. What do hams talk about, anyway? As in most casual contacts with people you don't know, warming up to a new contact takes a little time.

During the initial phase of a contact, both parties exchange information about the quality of the signals, their names, and their locations. This phase is a friendly way of judging whether conditions permit you to have an extended contact. Follow with information about your station and probably the local weather conditions. This information gives the other station an idea of your capabilities and indicates whether static or noise is likely to be a problem.

Common info exchanges

Here are the common items you exchange when making a contact:

- **Signal report:** This report is an indication of your signal's strength and clarity at the receiving station. You submit it as follows (see Table 8-5 for details):
  - *SSB:* A two-digit system communicates readability and strength, although sometimes, you can use a single Quality report from Q1 to Q5.
  - *CW and RTTY:* The same two-digit system is used for readability and strength, but a third digit is added to indicate the purity of the transmitter's note — rarely anything but 9 nowadays, because transmitting equipment is quite good. If you encounter a poor signal, however, don't hesitate to make an appropriate report.
  - *Digital modes:* It's becoming common to use a Readability, Strength, and Quality (RSQ) report, with the final digit reporting the quality of how a signal is being decoded. (For more information on RSQ, see `www.rsq-info.net`.)
  - *FM:* The signal report is the degree to which the noise is covered up, or *quieting*.
- **QTH (location):** On HF, where contacts take place over long distances, you generally give your town and state or province. You can give an actual address if you're asked for it, but if you aren't comfortable doing so, you don't have to. On VHF/UHF, you report the actual physical location, particularly if you're using a mobile radio.
- **Rig:** You can just report the power output shown on your transmitter's power meter (such as 25 watts) or give the full model number and let your contact assume that the transmitter is running at full output power.

- **Antenna:** Typically, you just report the style and number of elements, such as a two-element quad or ⅝-wave whip. Sometimes, you can report a specific model number.
- **Weather:** Remember that stations outside the United States report the temperature in degrees Celsius. Standard weather abbreviations that you can use for CW and digital modes include SNY, CLDY, OVRCST, RNY, and SNW. A Russian ham in Siberia once gave me a weather report of "VY SNW" (very snow).

Table 8-5 Reporting Signal Quality

Mode	*System*	*Report Definitions*
SSB	RS: Readability and Strength	R is a value from 1 to 5. The value 5 means easy to understand, and 3 means difficult to understand; 1 and 2 are rarely used.
		S is a value from 1 to 9. This number generally corresponds to the rig's signal-strength-meter reading on voice peaks.
	Quality number (*number*): Indicates overall quality	Q(*number*) is a value from Q1 to Q5. Q5 indicates excellent readability; reports below Q3 are rare.
CW	RST: Readability, Strength, and Tone	R is a value from 1 to 5; the values mean the same as for SSB.
		S is a value from 1 to 9; the values mean the same as for SSB.
		T is a value from 1 to 9. The value 9 is a pure tone, and 1 is raspy noise. The letter *C* is sometimes added to indicate a chirpy signal.
Digital (alternative)	RSQ: Readability, Strength, and Quality	R is a value from 1 to 5. The values mean the same as for SSB.
		S is a value from 1 to 9. The values mean the same as for SSB.
		Q is a value from 1 to 9, reflecting the modulation quality of your signal.
FM	Level of quieting (signal report is for the station calling, not the repeater's output signal strength)	*Full quieting* means that all noise is suppressed. *Scratchy* means that enough noise is present to disrupt understanding. *Flutter* means rapid variations in strength as a vehicle is moving. *Just making it* means that the signal is only strong enough to actuate the repeater and not good enough for a contact.

Further discussion

After you go through the first stages of the QSO, if the other ham wants to continue, you can try discussing some other personal information. The possible topics are almost endless — your age, your other hobbies, what you do for a living, your family members, any special interests, and ham-radio topics such as propagation conditions or particularly good contacts that you made recently.

The FCC forbids obscene speech (which is pretty rare on the air). The three topics that seem to lead to elevated blood pressures are politics, religion, and sex — hardly surprising. So hams tend to find other things to talk about. Oh, sure, you'll find some arguments on the air from time to time, just as you do in any other group of people. Don't be drawn into arguments yourself, however; no one benefits. Just "spin the big knob" and tune on by.

Try to keep your transmissions short enough that the other station has a chance to respond or that someone else can break in (see "Breaking in," earlier in this chapter). Also, if propagation is changing, or if the band is crowded or noisy, short transmissions allow you to ask for missed information. But you can have a QSO just as long and detailed as both parties want it to be.

At the conclusion of the contact, you might encourage the other station to call in again. Lifelong friendships are sometimes forged on the ham bands!

Calling CQ

After you make a few contacts, the lure of fame and fortune may become too strong to resist. It's time to call CQ yourself.

Anatomy of a CQ

A CQ consists of two basic parts, repeated in a cycle:

- **CQ itself:** The first part is the CQ itself. For a general-purpose "Hello, World!" message, just say "CQ." If you're looking for a specific area or type of caller, you must add that information, as in "CQ DX" or "CQ New England."
- **Your call sign:** The second part of the CQ is your call sign. You must speak or send clearly and correctly. Many stations mumble or rush through their call signs or send them differently each time, running the letters together. You've probably tuned past CQs like those.

A few CQs, followed by "from" or "DE," and a couple of call signs make up the CQ cycle. If you say "CQ" three times, followed by your call sign twice, that's a *3-by-2 call.* If you repeat that pattern four times, it's a *3-by-2-by-4 call.* At the end of the cycle, you say "Standing by for a call" (or "K"), to let everybody listening know that it's time to call.

Here's an example of a 3-by-3-by-3 on CW or a digital mode: "CQ CQ CQ DE NØAX NØAX NØAX CQ CQ CQ DE NØAX NØAX NØAX CQ CQ CQ DE NØAX NØAX NØAX K."

Depending on conditions, repeat the cycle (CQ and your call sign) two to five times, keeping it consistent throughout. If the band is busy, keep it short. If you're calling on a quiet band or for a specific target area, four or five cycles may be required. When you're done, listen for at least a few seconds before starting a new cycle to give anyone time to start transmitting.

CQ tips

Here are a few CQ do's and don'ts.

Do:

- Keep your two-part cycle short to hold the caller's interest.
- Use standard phonetics for your call sign (refer to Table 8-4, earlier in this chapter) on voice modes at least once per cycle.
- Send CW at a speed you feel comfortable receiving.
- Strive to sound friendly and enthusiastic.
- Wait long enough between CQs for callers to answer.

Don't:

- Mumble, rush, or slur your words.
- Send erratically or run letters together.
- Drag the CQ out. A 3-by-3-by-3 call is good for most conditions.
- Shout or turn up the microphone audio level too far. Clean audio sounds best.

Treat each CQ like a short advertisement for you and your station. It should make the listener think, "Yeah, I'd like to give this station a call!"

Saying the long goodbye

If hams do one thing well, it's saying goodbye. Hams use abbreviations, friendly names, phrases, and colloquialisms to pad their contacts before actually signing off. You rarely hear anyone say "Well, I don't have anything more to say. W1XYZ signing off." Sometimes, signing off takes as long as signing on.

Toward the end of the contact, let the other station know that you're out of gas. Following are some good endings:

- **I AM QRU:** In Morse code–speak, *QRU* means *out of things to talk about.*
- **See you down the log:** Encourage another contact at a later time.
- **BCNU:** Morse code for *be seein' you.*
- **CUL:** Morse code for *see you later.*
- **88 or Old Man:** Throw love and kisses to any female operator, whether you know her well or not. Or tack on an affectionate "Old Man" (if appropriate).
- **73:** Don't forget to send your best regards.
- **Pulling the big switch or GOING QRT:** If you're leaving the airwaves, be sure to say so after your call sign on the last transmission. On voice, say "NØAX closing down." On digital or Morse, send "NØAX SK CL." Anyone receiving these transmissions will know that you're vacating the frequency.

Chapter 9

Casual Operating

In This Chapter

- Using and linking repeaters
- Making a casual contact
- Copying Morse code
- Using e-mail over the air

After you tune around the amateur bands for a while, you'll agree that the lion's share of the ham's life is making relaxed, casual contacts. Some contacts are just random "Hello, anybody out there?" encounters. You'll also hear contacts between hams who are obviously old friends or family members who meet on the air on a regular basis.

It's not uncommon for hams to use ham radio to maintain contact, so to speak, while they're traveling off the grid. (I realize that this may be a shock, but in some places on Earth, a mobile phone's service indicator won't light up.) In these remote places, ham radio is available to fill the breach. It's fun, and people enjoy contacting travelers wherever they are.

I cover the technical aspects of station configuration and operation in Part IV. (For more technical stuff, see the *Ham Radio For Dummies* website at `www.dummies.com/extras/hamradio`.)

In this chapter, you find out about the different ways to conduct these relaxing contacts. As with most things in life, "There's kindy a knack to it," as my dear Aunt Lexie used to say.

Before I start on this operating business, allow me to suggest that you get two books:

- ***The FCC Rules and Regulations for the Amateur Radio Service:*** This book is available from the ARRL (`www.arrl.org/shop`) for only a few dollars. It conveniently includes not only the rules themselves, but also a clear discussion of do's and don'ts, along with information on technical standards and the FCC Universal Licensing System. Hams really should have a copy in their shack, whether they're veterans or beginners.

- ***ARRL Operating Manual:*** The manual dedicates a separate chapter to all kinds of on-the-air operating; provides handy references, tables, and maps; and answers just about any operating question you can come up with. It's also available at `www.arrl.org/shop`.

Operating on FM and Repeaters

Most new hams begin operating as Technician class licensees, with access to the entire amateur VHF and UHF bands. By far the most common means of communicating on those bands is through the use of an FM repeater (introduced in Chapter 8).

Finding a repeater

Figure 9-1 explains the general idea behind a repeater. A repeater receives FM signals on one frequency and simultaneously retransmits (or *repeats)* them on another frequency. The received signals aren't stored and played back; they're retransmitted on a different frequency at the same time they're received, in a process called *duplex operation.*

Repeaters use FM instead of single sideband (SSB) because of the relative simplicity of the transmitters and receivers — an important consideration for equipment that's operating all the time and needs to be reliable. FM is also relatively immune from static if signal strength is good, so it makes for a more pleasant contact. Except for a small segment of the 10 meter band, FM is rarely found on HF due to restrictions on signal bandwidth and to FM's relatively poor quality for weak signals when compared with SSB. Above 30 MHz, FM's qualities are ideal for local and regional coverage.

If the repeater is located on a high building, tower, or hill, its sensitive receiver picks up signals clearly from even tiny handheld radios. Then it uses a powerful transmitter to relay that input signal over a wide area. Stations can be separated by tens of miles yet communicate with a watt or two of power by using a repeater.

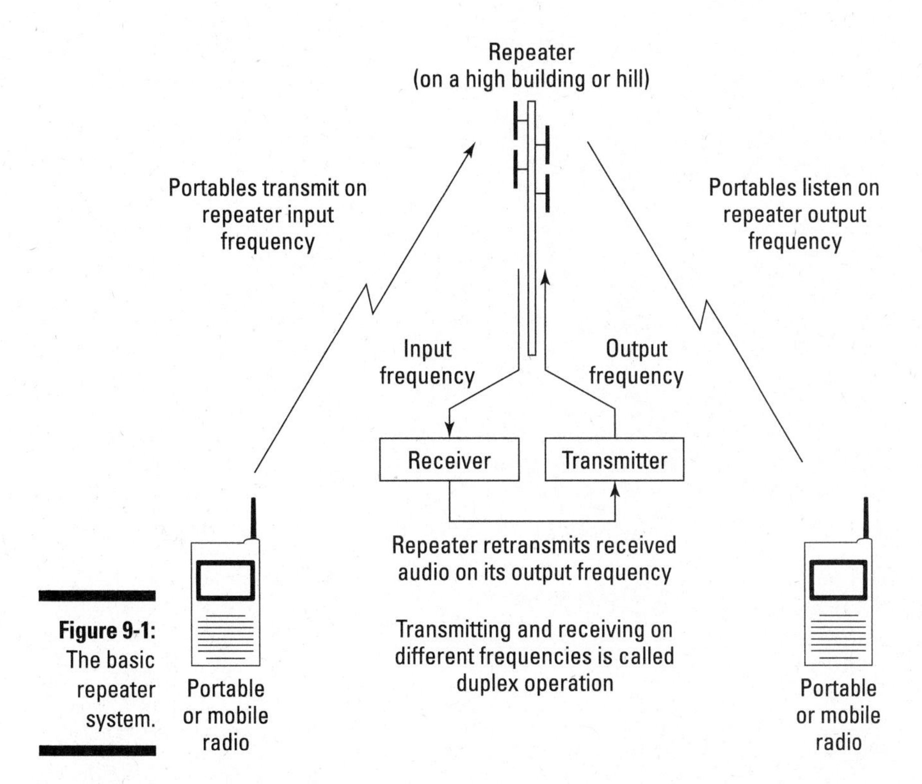
Repeater
(on a high building or hill)
Portables transmit on repeater input frequency
Portables listen on repeater output frequency
Input frequency
Output frequency
Receiver
Transmitter
Repeater retransmits received audio on its output frequency
Transmitting and receiving on different frequencies is called duplex operation
Portable or mobile radio
Portable or mobile radio

Figure 9-1: The basic repeater system.

TIP

Ham radio repeaters are constructed and maintained by radio clubs or private citizens as a service to the ham community. Installing equipment on towers and high buildings often involves rental or access fees, even for not-for-profit amateur groups. If a repeater users' group or club exists in your area, consider joining or donating to it to help defray the cost of keeping the repeater on the air.

Understanding repeater frequencies

To make repeater communications work, you have to know the frequency on which a repeater listens and the frequency on which it's transmitting. The listening frequency (the one that listens for your signal) is called the repeater's *input* frequency, and the frequency that you listen to is called the repeater's *output* frequency. The difference between the two frequencies is called the repeater's *separation* or *offset.* The combination of a repeater's input and output frequencies is called a *repeater pair.*

As Figure 9-2 shows, repeater pairs are organized in groups, with their inputs in one part of the band and their outputs in another, all of them having a common separation. Each pair leapfrogs its neighbor, with each input or output channel separated by the same frequency span, called the *channel spacing*. The input may be a lower frequency than the output, or vice versa.

Figure 9-3 shows the locations of repeater segments on the five primary VHF/UHF bands. The 6 meter band has three groups of repeaters: 51.12 to 51.98 MHz, 52 to 53 MHz, and 53 to 54 MHz. The 2 meter band also has three groups: 144.6 to 145.5 MHz, 146.01 to 147 MHz, and 147 to 147.99 MHz. The 70 cm band hosts one segment with a large simplex segment in the middle: from 440 to 449.99 MHz. The 222 and 1296 MHz bands have a single group. Repeaters are allowed on the 902 MHz and 2304 MHz bands but aren't common. If you have a license with HF privileges, you may want to give the 10 meter FM repeaters a try. They have output frequencies between 29.610 and 29.700 MHz and an offset of –100 kHz.

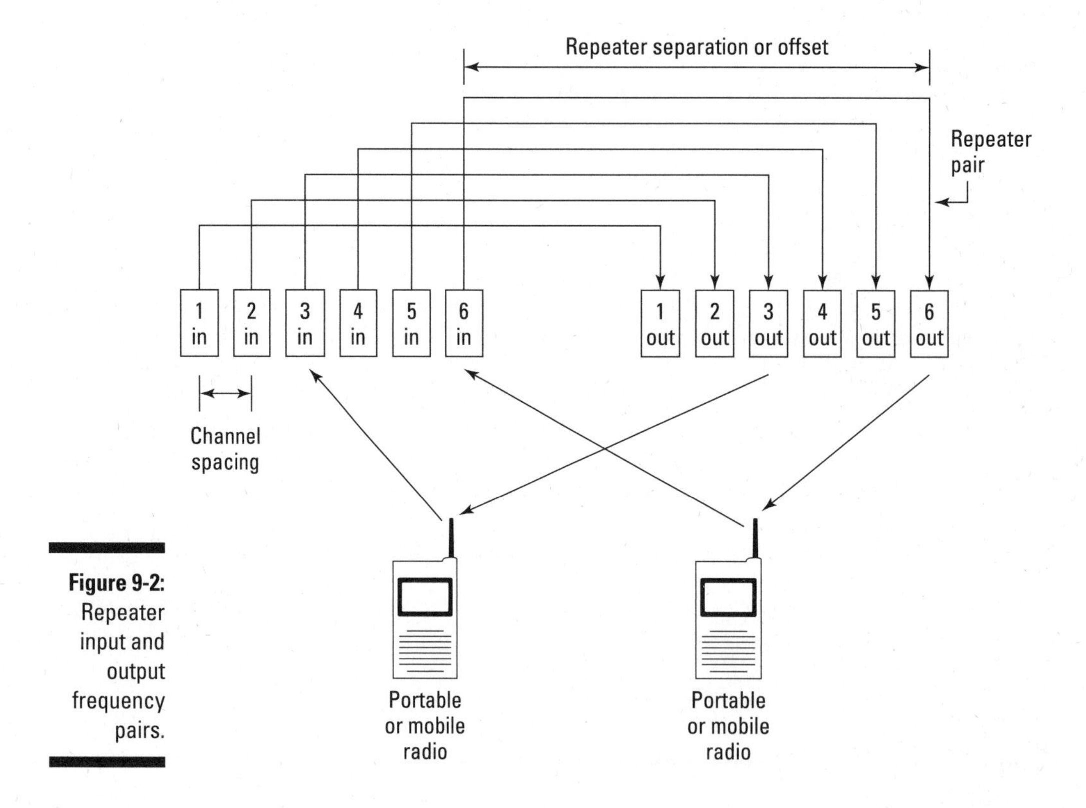

Figure 9-2: Repeater input and output frequency pairs.

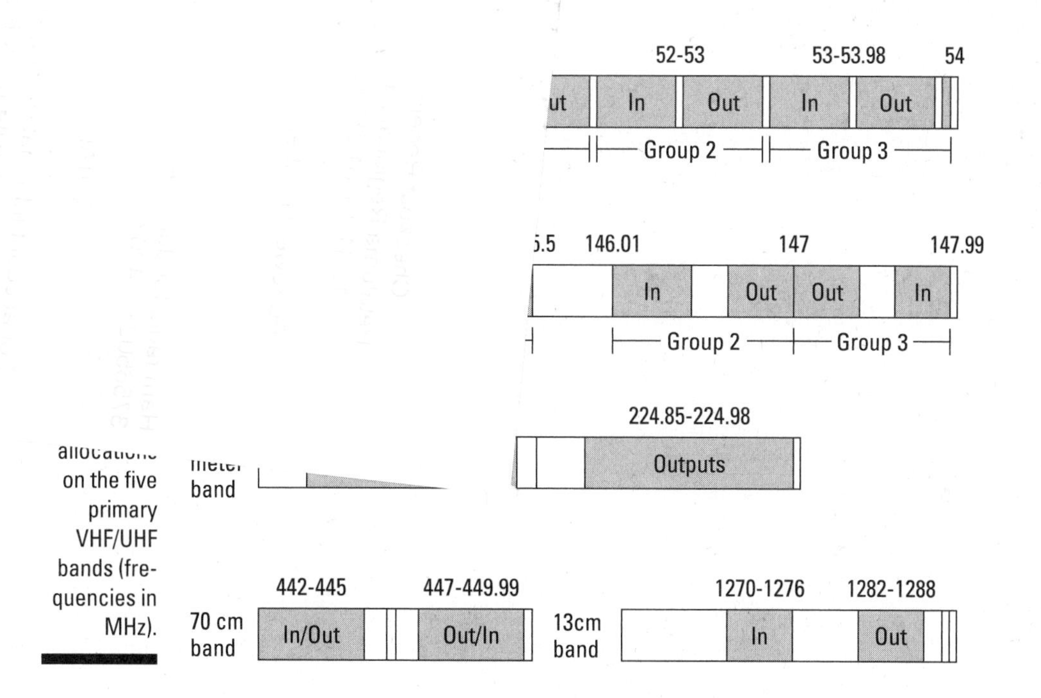

Finding repeater frequencies

Not all channels are occupied in every area. Also, around the country, some local variations exist in channel spacing and, in rare cases, offset. To find out where the repeater inputs and outputs are for a specific area, you need a repeater directory (see Chapter 8).

The online directory at `www.artscipub.com/repeaters/welcome.html` is a good resource for locating nearby repeaters. It also provides information about whether the repeater can be linked to other systems, as discussed later in this chapter.

If you don't have a repeater directory and are just tuning across the band, try using Table 9-1, which lists the most common output frequencies and repeater offsets to try. Tune to different output frequencies, and listen for activity.

Most FM VHF/UHF radios are capable of *scanning* so they can switch among channels rapidly, stopping when they receive a signal. Scanning allows you to monitor several repeaters for activity without having to switch channels manually.

Set your radio's offset appropriately for each band, and tune to the center frequency of each channel to avoid signal distortion from being off-frequency.

Table 9-1 Repeater Channel Spacings and Offsets

Band	*Output Frequencies of Each Group (In MHz)*	*Offset from Output to Input Frequency*
6 meters	51.62–51.98	–500 kHz
	52.5–52.98	
	53.5–53.98	
2 meters (a mix of 20 and 15 kHz channel spacing)	145.2–145.5	–600 kHz
	146.61–147.00	–600 kHz
	147.00–147.39	+600 kHz
220 MHz	223.85–224.98	–1.6 MHz
440 MHz (local options determine whether inputs are above or below outputs)	442–445 (California repeaters start at 440 MHz)	+5 MHz
	447–450	–5 MHz
1296 MHz	1282–1288	–12 MHz

Using repeater access tones

To minimize interference from other repeaters and strong nearby signals, extra audio tones are used as described in the following paragraphs. Sometimes, individual hams use these tones when talking to each other in crowded areas, such at hamfests or conventions. Find out what system is used and how to configure your radio to use it. Repeater directories provide the information you need.

Most repeaters now use a system called *tone access* — also known as subaudible, PL, or CTCSS (Continuous Tone Coded Squelch System). You may have used tone access on the popular Family Radio Service (FRS) and General Mobile Radio Service (GMRS) radios. where the tones are known as *privacy codes*. Tone access keeps a repeater or radio output quiet (or *squelched)* for all signals except those that carry the proper tone.

Using tone access

Regardless of what it's called, tone access works this way: When you transmit to a repeater, a low-frequency tone (between 69 Hz and 255 Hz) is added to your voice. (You can find a list of these tones and how to select them in your radio's operating manual.) When the repeater receives your transmission, it checks your voice for the correct tone. If it detects the correct tone, the repeater forwards your voice to the repeater output. This system prevents interfering signals from activating the repeater transmitter; these signals won't carry the correct tone signal and aren't retransmitted.

For a more detailed discussion of tone access, see the *ARRL Handbook* or the article "Decoding the Secrets of CTCSS," by Ken Collier (KO6UX), at `olyham.net/library/CTSSSecrets.pdf`.

Most radios also offer *tone squelch,* which uses the same set of tones to control the squelch circuit. It works just like the repeater receiver's tone access, in that only signals with the right tone are output as audio. Most of the time, the same tone is used for both transmitting and receiving.

Many recent radio models also have a tone-decoder function that detects which tone a repeater is using. If your radio doesn't have this function, and you don't know the correct tone, you can't use the repeater. How can you find out what the proper tone is? Check a repeater directory, which lists the tone and other vital statistics about the repeater. When you determine the correct tone, either via the tone-decoder function or the repeater directory, you can program your radio to send the correct tone and activate the repeater.

Using Digital Coded Squelch (DCS)

DCS is another method of reducing interference. It allows you to hear only audio transmitted by selected stations. DCS consists of a continuous sequence of low-frequency tones that accompanies the transmitted voice. If your receiver is set to the same code sequence, it passes the audio to the radio's speaker. If the transmission uses a different code, your radio remains silent.

Most people use DCS to keep from having to listen to all the chatter on a repeater, hearing only the audio of others who use the same DCS sequence, such as friends or club members.

Not all repeaters pass the tone access or DCS tones through to the transmitter and may filter them out.

Using simplex

When one station calls another without the aid of a repeater, both stations listen and talk on the same frequency, just as contacts are made on the HF bands, which is called *simplex* operation. Hams usually use FM simplex when they're just making a local contact over a few miles and don't need to use a repeater. Interspersed with the repeater frequency bands shown in Figure 9-1, earlier in this chapter, are small sets of channels designated for simplex operation.

If the station you're contacting via a repeater is nearby, it's good manners to switch to a simplex frequency instead of tying up the repeater.

Having a common simplex channel is a good way for a local group of hams to keep in touch. Simplex frequencies are usually less busy than repeater frequencies and have a smaller coverage area, which makes them useful as local or town intercoms. Clubs and informal groups often decide to keep their radios tuned to a certain simplex frequency just for this purpose. If these groups aren't having a meeting or conducting some other business, feel free to make a short call (such as "NØAX monitoring" or "K9SD, this is NØAX") and make a friend.

On bands that have a lot of space, such as the VHF and UHF bands, making contacts outside the repeater channels is easier if you know approximately where the other hams are. That's the purpose of calling frequencies: to get contacts started. You hear hams call CQ (the general "Come in, anybody" call) on a calling frequency and, after establishing a contact, move to a nearby frequency to complete it. If I call CQ on the 6 meter FM simplex calling frequency (52.525 MHz), and N6TR answers me, I say, "N6TR from NØAX. Hi, Larry. Let's move to five-four, OK?" This transaction means that I receive N6TR and am tuning to 52.54 MHz, a nearby simplex frequency.

Making a couple of complete contacts on calling frequencies is okay if the band isn't busy. Otherwise, move to a nearby frequency.

A national FM simplex calling frequency is set aside on each band just for general "Anybody want to chat?" calls. These frequencies are 52.525, 146.52, 223.50, 446.00, and 1294.5 MHz. When driving long distances, I often tune my radio to one of these channels to meet up with other travelers on the highways. Travelers may tune to these frequencies as well, so monitoring them by using your radio's scanning feature is a good way to make them feel welcome or to give directions. In recent years, GPS navigation has really cut down on "Can anybody help me get to . . . ?" calls, though.

If you're traveling and want to make a contact on the simplex calling frequencies, the best way to do so is to just announce that you're monitoring, as you would on a repeater channel, or make a transmission similar to this one:

> This is NØAX November Zero Alfa X-ray mobile, headed southwest on Interstate 44 near Leasburg. Anybody around?

Repeat this transmission a few times, spacing the repeats a few seconds apart. If you're moving, try making a call once every five minutes or so.

Because simplex communications don't take advantage of a repeater's lofty position and powerful signal, you may have to listen harder than usual on these frequencies. Keep your squelch setting just above the noise level. When you're making a call, you may want to open the squelch completely so that you can hear a weak station responding.

Solid simplex communications usually require more power and better antennas than typical handheld radios have — at least, on one end of the contact. To get better results on simplex with just a few watts, try replacing a handheld radio's flexible antenna with a full-size ground plane or a small beam. I discuss these antennas in Chapter 12.

Setting up your radio

After you figure out offsets, tones, and repeater frequencies, take a few minutes to check out your radio's operating manual. To use your radio effectively, you need to know how to do each of the following things:

- **Set the radio's receive frequency and transmit offset.** Know how to switch to simplex (no offset) or to listen on the repeater's input frequency. Some radios have a REV (reverse) button just for this purpose.
- **Switch between VFO and Memory modes.** In VFO mode, a radio can be tuned to any frequency; this is usually how you select a frequency to be stored in a memory channel. In Memory mode, tuning the radio changes from channel to channel.
- **Turn subaudible tones on or off and change the tone.** If your radio can detect and display the tone frequency being used (a feature called *tone scanning)*, know how to use it.
- **Control the Digital Squelch System (DCS) function.** You'll need to know to activate and deactivate DCS in your radio. To use DCS, you also need to know how to pick the tones used to make the DCS code.
- **Store the radio settings in a memory channel, and select different memory channels.** Storing information in a memory channel (called *programming a channel)* usually requires you to use the VFO mode to configure the radio just the way you like it. Then you press a "memory write" button, select the channel number you want, and press the memory write button again to program the radio's memory for that channel. Some radios have hundreds of programmable memories.

Making contacts, FM style

Because VHF and UHF FM voice contacts are usually local or regional, they tend to be used to connect with personal contacts rather than make random acquaintances. Most hams use a few favorite repeaters or simplex frequencies as a sort of regional intercom. They turn on a radio in the shack at home or in the car and monitor a channel or two to keep an ear out for club members or friends. Even though several people may be monitoring a repeater, these hams mostly just listen unless someone calls them specifically or they hear a request for information or help.

This style can be a little off-putting to hams who are new to FM and can even seem unfriendly at times. Rest assured that the hams aren't being unfriendly; they're just not in "meet and greet" mode on the repeater or a favorite simplex channel. Imagine the difference between meeting someone at a party versus at a grocery store. At the party, everyone expects to make new acquaintances and has conversations. At the store, people aren't there as a social exercise and may even seem to be a little brusque. With this idea in mind, the following sections provide some suggestions to help you get comfortable with the FM style.

Joining the group

The best way to become acquainted with a group is to participate in its activities. In areas with good repeater coverage — such as cities and suburban areas — nets are a very common group use of FM repeaters, for example. The most common nets on FM are for emergency-services groups, weather and traffic (the automobile kind), and equipment swap meets. Technical assistance or question-and-answer nets are common in the evenings and on weekends. Use the ARRL net-search page at `www.arrl.org/arrl-net-directory-search` to find local nets on the VHF and UHF bands. Your Elmer (see Chapter 3) may be able to help you with times and frequencies, and so will other radio-club members.

Almost all nets call for visitors, generally at the end of the net session, and that's your chance. When you check in (by giving your call sign and maybe your name; follow the given directions), ask for an after-net contact with the net control station or a station that said something of interest to you. After-net contacts are initiated on the net frequency when the net is completed. Sometimes, they're held on the net frequency; at other times, the stations establish contact there and then move to a different frequency.

During the after-net contact (QSO), you can introduce yourself and ask for help finding other nets in the area. If you have specific interests, ask whether the station knows about other nets on similar topics. Ideally, you'll get a referral and maybe even a couple of call signs to contact for information.

After you check in to a few nets, your call sign starts to become familiar, and you have a new set of friends. If you can contribute to a weather or traffic net, or deliver a message destined for your town or neighborhood, by all means do so. Contributing your time and talents helps you become part of the on-the-air community in no time, and it's good operating practice too.

Seizing the opportunity

As you monitor the different channels, you quickly discover which repeaters encourage conversations and which don't. If you can identify a repeater that's ragchew-friendly, you'll have a fairly easy time making a few casual contacts. Listen for a station accessing the repeater, which sounds something like this: "NØAX monitoring" or "NØAX for a call." When you hear that transmission, respond with a quick 1-by-2 call using phonetics, such as "NØAX, this is KD7DQO, Kilo Delta Seven Delta Quebec Oscar, over." By convention, calling CQ is reserved for SSB and Morse code (CW) operations when you're not sure who's out there and signals are generally weak. To fit in with FM's strong-signal intercom style, make short transmissions.

Continue your contact as you do other contacts (see Chapter 8), but keeping your transmissions short is important. Repeaters have a time-out function that disables the transmitter if the input is busy for more than just a few minutes (typically, three). This function prevents the transmitter from overheating and also keeps a long-winded speaker from locking out other users. When you're transmitting, if you let up on the mike switch and no repeater signal comes back, you may have timed out the repeater. Just let the repeater rest for 10 or 15 seconds until the receiver is re-enabled and the timer resets. There's no need to be embarrassed about timing out a repeater (unless you keep doing it), because it's a rare ham who hasn't done it.

Open and closed repeaters

If you purchase a repeater directory, you may see some repeaters that are marked as *closed,* meaning that they're not open to the ham-radio public. Some repeaters are closed because they're dedicated to a specific purpose, such as emergency communications. Other repeaters are intended to be used only by members of their supporting group.

Rest assured that you can use a closed repeater in case of emergency, but respect the wishes of its owners, and don't use it for casual operation. If you aren't sure whether a repeater is closed or not, transmit to it and say something like this: "This is NØAX. Is a control operator monitoring?" If you get a response, that person is the one to ask.

Getting technical about repeaters

Repeaters are widely used all over the world, and a wealth of other information is available for you to consult as you grow more experienced and curious about repeater and FM operations. Check the following websites for more information:

- **AC6V website** (`www.ac6v.com/repeaters.htm`): This site offers comprehensive links to many ham radio topics.
- **eHam.net:** This ham radio portal includes numerous areas of interest to hams, including a handy Guide to Amateur Radio for New Hams page (`www.eham.net/newham`). Click the Basic Operating link for information about repeater operating.
- **ARRL Technical Information Service** (`www.arrl.org/technical-information-service`): This site has many public links and numerous in-depth articles for ARRL members.

Recognizing Repeater Features

You can find an amazing set of features in repeater land. Many repeaters have voice synthesizers that identify the repeater and announce the time and temperature. Hams who use the repeater can activate and deactivate some functions, such as autopatch and automated announcements of time or temperature, by using telephone dialing tones on the keypads commonly used on microphones. Repeaters are linked to provide wide coverage even across bands or may have connections via the Internet, as described in "Repeater networks," later in this chapter.

In the next few sections, I touch on the basics of repeater operation and features.

Autopatch

Another feature of repeaters is called *autopatch,* which allows a repeater user to make a telephone call through the repeater. When you access a repeater's autopatch function, a dial tone appears on the air. Then dial the number, using the numeric keypad on your radio or microphone. The tones feed through the telephone system, which dials the number and connects you. Even in a world full of mobile phones, autopatch can still come in handy if your mobile phone's battery goes out or the service network malfunctions.

All autopatch calls occur over the air; they're not private.

Respecting the limits of autopatch

Although you can do a lot of important things with autopatch — most hams use it to call 911 to report accidents, for example — you should use this function wisely. Also, you must use special access codes to activate the feature. To get these codes, you may be required to join the club or group that maintains the repeater. Some limits exist as to what you're allowed to do via autopatch. Conducting business is forbidden, for example, as it is on any amateur frequency. (You can perform a limited amount of personal business, however, such as calling a store or — my favorite — ordering a pizza.) Finally, most repeater systems place strict limits on the numbers you can dial so that they don't incur long-distance charges.

In certain conditions, such as when your cellphone can't find a suitable service provider or the mobile phone systems are overloaded or unreachable, autopatch may be able to get through. This access is particularly important during disasters and emergencies.

Making a phone call with autopatch

If your radio has the autopatch feature, and you obtain the tone sequences, follow these general steps to make a phone call (the exact steps may vary from repeater to repeater):

1. **In an emergency, break into an ongoing contact, and ask the stations to stand by for an emergency autopatch.**

   If there's no emergency, wait until the channel is clear.

2. **Announce that you're switching to autopatch.**

3. **Holding down the microphone button, press the activation sequence of tones; then release the mike button.**

   You might say "NØAX accessing autopatch" and press *82 or #11, for example.

   The repeater acknowledges your sequence with some kind of tone, a synthesized voice, or maybe the telephone-system dial tone.

   If you don't hear an acknowledging transmission, go back to Step 2. You may have trouble if your signal is fluttery or weak.

   If necessary, another repeater user can activate the autopatch function for you.

4. **When you hear the dial tone, press the tone keys for the number you want to dial.**

You may have to precede the telephone number with a dialing code, depending on the repeater's specific operating requirements.

5. **When someone answers your call, immediately state that you're using a radio.**

   You can say, "This is *(your name),* and I'm calling you via amateur radio, so this call is not private." This transmission lets the other person know that a radio link is involved and that both of you can't talk at the same time. Emergency-services dispatch operators generally are familiar with autopatch, but pizza-delivery operators may not be.

6. **Transmit your message.**

   Keep it short and appropriate for a public conversation.

7. **When you finish, hold down the mike button, key in the hangup tones (if needed), and announce that you're disabling autopatch.**

   You might say, "This is NØAX releasing autopatch."

   The repeater acknowledges the release with an announcement similar to the one it made at activation.

8. **(Optional) Let the other repeater users know that you're finished and that they can resume normal operation.**

   Don't forget to thank them for standing by or for providing any assistance they may have given you.

Repeater networks

Within a local or regional area, many repeater systems may use *remote receivers* that relay weak signals from outlying areas back to the main repeater transmitter. The relayed signal is transmitted over a *control link,* which is a dedicated transmitter and receiver operating on a VHF or UHF band. It's not unusual for a repeater to have three or more remote receivers. To link repeaters over wide areas and long distances, however, it is common for repeater systems to use the Internet.

Hams have three primary ways to link their repeaters via the Internet: EchoLink (`www.echolink.org`), D-STAR (`www.dstarinfo.com`), and IRLP (`www.irlp.net`). EchoLink and IRLP use Voice over Internet Protocol (VoIP) to share voice signals between repeater systems, in much the same way that the Skype web communication application does. D-STAR uses its own set of protocols to link individual and multiple repeaters through gateways. Figure 9-4 illustrates the basic IRLP, and Figure 9-5 shows D-STAR.

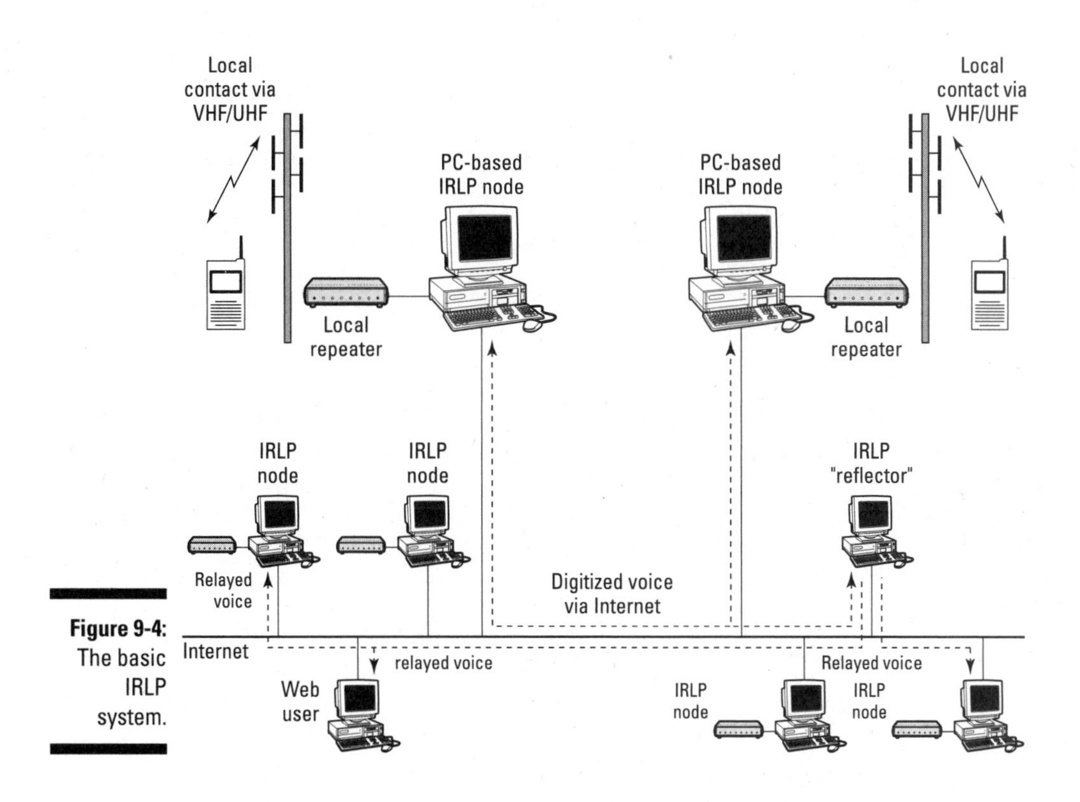

Figure 9-4: The basic IRLP system.

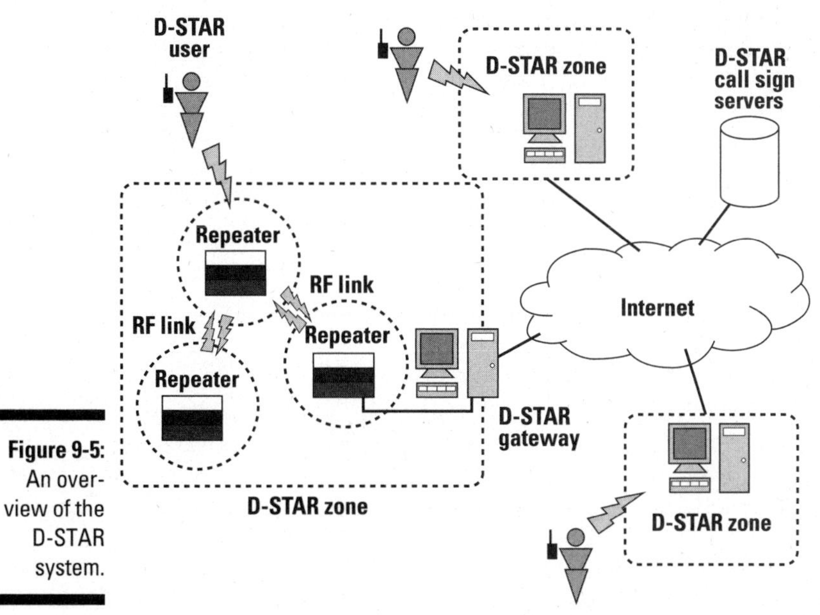

Figure 9-5: An overview of the D-STAR system.

EchoLink and IRLP systems can be used by hams with any kind of FM voice radio. D-STAR's digital voice protocol must be used by everyone on the system, requiring the use of D-STAR radios. D-STAR's fully digital system also enables data-only operation, and many software applications have been developed to use that capability.

Complete information about how the repeater networks are constructed and used is available on the home pages for each system, along with extensive "getting started" instructions, FAQs, and resources for finding and even installing your own system repeater.

All three systems — IRLP, EchoLink, and D-STAR — are active, and more repeaters are linked to the systems every day. The IRLP system includes about 3,000 stations around the world. EchoLink currently lists more than 2,000 repeaters and more than 200 conference servers. The D-STAR system lists about 1,000 gateways, many serving more than one repeater. To find other hams who use these systems, you'll want to join one of the online user groups:

- **EchoLink:** `http://groups.yahoo.com/group/echolink`
- **D-STAR:** `www.d-starusers.org`
- **IRLP:** `http://groups.yahoo.com/group/irlp`

Using EchoLink and IRLP

The links between repeaters and individual stations in the IRLP and EchoLink systems are controlled manually by the system users. When you're connecting to either system beyond your local repeater, you must enter an access code manually (a bit like using a repeater's autopatch feature, as described earlier in this chapter). The code identifies the repeater system you want to use; then the system sets up the connection and routes the audio for you. When you finish, another code or a disconnect message ends the sharing. This overview is very simplified, of course, and both systems offer useful features beyond simple voice links.

EchoLink systems can also be accessed directly from a computer via the Internet, so users don't have to have a radio at all, making EchoLink a popular way to communicate if you don't have a radio handy or are traveling away from your home repeater.

Using D-STAR

D-STAR is not only a type of repeater system, but also a complete set of digital communication protocols for individual radios. A station that uses a D-STAR radio communicates by sending a stream of digitized audio over the air to another D-STAR radio or a D-STAR repeater. Regular FM voice radios can't receive D-STAR signals, and vice versa. (Some radios can operate in either FM or D-STAR mode.) A D-STAR repeater receives the digital stream of data and retransmits it just like an FM repeater does, but still as digital data. Because the voice signal is already digital when it's transmitted, it can be shared or relayed over digital networks easily.

D-STAR repeaters are connected to the D-STAR system through *gateways:* computers that are connected to the repeater and to the Internet. Hams establish links from one D-STAR repeater to another by entering call signs into their D-STAR radios. The D-STAR system's servers use these call signs to look up the low-level Internet network addresses of individual repeaters. Then the system directs each repeater to make the connection and share the voice data.

Call-sign lookup is an interesting feature because hams don't have to know where another ham is to contact that person; the D-STAR system remembers where each ham last used a D-STAR repeater. Calls to a specific ham are routed to his or her last-known D-STAR repeater. This feature is a lot like the mobile phone system, which keeps track of where your phone is connected to the network so it can route your calls to that point.

Using an IRLP reflector

An IRLP *node* is a regular FM repeater with an Internet link for relaying digitized voice. A user or control operator can direct an IRLP node to connect to any other IRLP node. When the node-to-node connection is made, the audio on the two repeaters is exchanged, just as though both users were talking on the same repeater. It's common for a ham in Europe to communicate with a ham in New Zealand, for example, with both parties using handheld radios that put out just a watt or two.

You can also connect several nodes by using an IRLP reflector. The reflector exchanges digitized audio data from any node with several other nodes in real time. Even a user who doesn't have a radio can join in by logging on to an IRLP reflector or node. (All users who create radio transmissions have to be licensed, however.)

Using the IRLP system is very much like using an autopatch system. You don't need anything more than your radio, the IRLP system's control-tone sequences for your repeater, and a list of the four-digit node on-codes that form the IRLP address of an active IRLP repeater.

To connect to another IRLP-enabled repeater, follow these steps:

1. **Enter an IRLP access code.**

   The access code sets up the repeater to accept an IRLP on-code. This process is just like activating autopatch.

2. **When the repeater indicates that the IRLP system is ready, enter the tones that send the on-code of the repeater that you want to connect to.**

   Entering the tone is just like entering a telephone number into an autopatch system. You can find a list of available IRLP nodes and their on-codes at `http://status.irlp.net`.

   After a short delay, the node you connected to identifies itself in plain voice with a call sign and location, and you're connected.

3. **Make your contact.**

   Any transmissions that you make are retransmitted on the remote node, and you also hear all the audio from the other node. If the other node is busy with another IRLP connection, you hear a message to that effect. Try another node, or come back later.

4. **When you finish, use the IRLP control codes to disconnect, as with autopatch.**

You can get more complete information on the IRLP system through several web resources. An introductory article at `www.eham.net/newham/irlp` gives a good overview of the system. If you're really interested in the complete technical details and maybe even in putting together your own IRLP node, `www.irlp.net` has complete details.

Chewing the Rag

I first mention ragchewing in Chapter 8 as a likely way to make your first contacts. In this section, I go into deeper detail about the etiquette of the ragchew.

Knowing where to chew

You know enough about ham radio to realize that you can't just spin the dial and start bellowing into a microphone or pounding out characters on a keyboard or Morse key. Although ragchewing isn't listed on any band plan, you can find ragchewers on certain parts of every band.

HF bands

Below 30 MHz, all the bands have a similar structure. CW (Morse code) and digital modes occupy the lower third (more or less), and voice modes occupy the upper two-thirds (less or more).

Figure 9-6 shows how the general operating styles are organized on a typical HF band. You find the ragchewers mixed in with the long-distance (DX) contacts at the low end of the band.

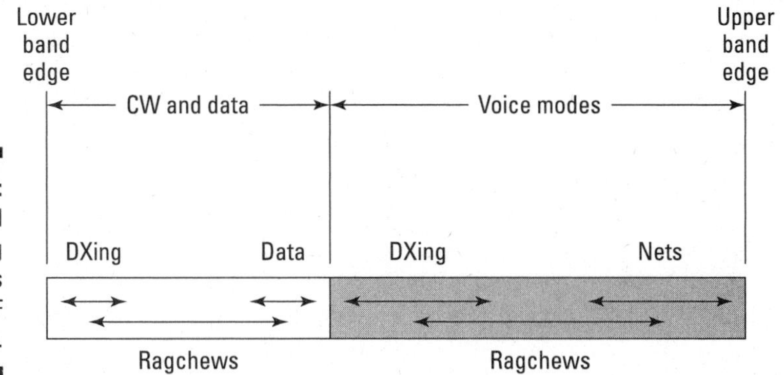

Figure 9-6: The general operating conventions on the HF bands.

You can always find code at the low frequencies of the HF bands. The faster operators tend to be at the bottom of the bands, with average code speed dropping slowly as you tune higher.

As you start, find an operator sending code at a speed you feel comfortable receiving. Slow-speed code operators are at the high end of the General sub-bands — around 3.600, 7.100 to 7.125, 21.100 to 21.175, and around 28.100 MHz. Medium-speed QSOs are the norm elsewhere, even mixed in with high-speed operators down low in the band.

Digital signals nearly always cluster close to the published calling frequencies listed in the band plans for that particular mode unless a major contest or some other event is going on. Stations spread out higher in frequency from there.

You may think that ragchewers are buffeted from all sides, but that's not really the case. Ragchew contacts take place all the time, so they tend to occupy just about any spare bit of band. To be sure, in the case of emergencies (when a lot of nets are active), major expeditions to rare places, or weekends of big contests, the bands may seem to be too full for you to get a word in edgewise. In those cases, try a different band or mode; you'll probably find plenty of room.

The FCC can declare a communications emergency and designate certain frequencies for emergency traffic and other communications. Keeping those frequencies clear is every amateur's responsibility. The ARRL transmits special bulletins over the air on W1AW, by e-mail, and on its website if the FCC does make such a declaration. The restrictions are in place until the FCC lifts them.

VHF and UHF bands

On the VHF and UHF bands, you usually find ragchewing in the repeater sections, although wide-open spaces for an SSB, CW, or digital conversation are available in the so-called "weak signal" portions. Table 9-2 lists the calling frequencies and portions of the VHF/UHF bands. The operating style in this portion of the bands is similar to HF as far as calling CQ and making random contacts go, but the bands are far less crowded because propagation generally limits activity to regional contacts.

Table 9-2 VHF/UHF CW and SSB Calling Frequencies

Band	Frequencies (In MHz)	Use
6 meters	50.0–50.3	CW and SSB
	50.070, 50.090	CW calling frequencies
	50.125 and 50.200	SSB calling frequency; use upper sideband (USB)
2 meters	144.0–144.3	CW and SSB
	144.200	CW and SSB calling frequency; use USB
222 MHz (1¼ meters)	222.0–222.15	CW and SSB
	222.100	CW and SSB calling frequency; use USB
440 MHz (70 cm)	432.07–433.0	CW and SSB
	432.100	CW and SSB calling frequency; use USB

The bottom portion of the VHF/UHF bands is referred to as *weak signal,* although that's really a misnomer. The reason for the name is that contacts via CW and SSB can be made with considerably weaker signals than on FM. Digital contacts on modes like PSK31 are even better for solid copy on very weak signals. Most of the weak-signal signals you hear are sufficiently strong for excellent contacts, thank you!

Knowing when to chew

Whether you're on HF or VHF/UHF, you'll find that ragchewing has its good times and its poor times. When calling CQ (signifying that you want to talk to any station), you can let it be known in several ways that you're looking for an extended contact. You also hear numerous clues that a ragchew may not be what another station has in mind.

HF-band DXing

DXing takes place at the low end of the HF bands for two reasons:

- **Long-distance contacts generally are more difficult to make than regional or local contacts are.** The signals are weaker, and it's easier for DX enthusiasts to find one another if they tend to operate in one area of the band.
- **After the bands were divided into different sub-bands for the different license classes, fewer operators were able to use the lowest segments of the bands.** DX stations took advantage of less crowding in those segments.

Sometimes, if a super-rare station comes on the air, the sheer numbers of DXers calling (a *pileup*) crowd out sustained contacts, so the ragchewers move up the band. The ragchewers space themselves out around the nets, round-tables, calling frequencies, and data signals, taking advantage of ham radio's unique frequency agility to find an empty spot.

Because DXing tends to attract a crowd, this concept is somewhat incompatible with the more ordered operating style of nets. Therefore, those types of operations would gather at the other end of the bands. With this structure, groups can engage in their preferred style of operating without interference. Data and Morse code signals tend to be incompatible, so the digital signals, such as radioteletype (RTTY) and PSK, also stay toward the high portion of the band. The popular 40 meter band is an exception, as the limited amount of space that hams share worldwide here results in digital mode operation moving lower into the traditionally CW part of the band. We all have to share!

I keep coming back to this point, but listening is the best way to learn operating procedures in ham radio. The most important part of any amateur's station is between the ears. If you want to call CQ successfully, spend some time listening to more experienced hams do it.

Good times

Assuming that you're tuning an open band (with signals coming in from various points), when is a good time to ragchew? First, consider the social aspects of your contact timing. Weekdays generally are good days to ragchew, especially during the daylight hours, when hams who have day jobs are at work or younger hams are in class.

It may be a different time of day at the other end of the contact, particularly if the other station is DX.

You may want to revisit Chapter 8 when considering what band is best to use. If you like to talk regionally, you can always use a repeater or one of the low-frequency HF bands. For a coast-to-coast talk, one of the high-frequency bands is your best bet. The better your antenna system, the more options you have.

What's a WARC?

Hams refer to 30, 17, and 12 meters as the WARC bands. *WARC* refers to an international World Administrative Radio Conference. At the 1979 conference, amateurs worldwide were granted access to three new bands: 30, 17, and 12 meters. These new bands were immediately nicknamed the WARC bands to distinguish them from the older ham bands at 160, 80, 40, 20, 15, and 10 meters. The 60 meter band was opened to amateurs in 2002 on a limited basis. Although hams didn't gain access to it through the WARC process, it's still lumped in with the true WARC bands as not being one of the traditional bands.

By convention, contest operation doesn't take place on the WARC bands and 60 meters. If you prefer not to deal with contest QRM (interference), these bands are good choices.

Lots of hams do their operating on weekends, but that's also when special events and contests are held. Be prepared for a full band every weekend of the year. The silver lining of this cloud is that plenty of hams will be on the air for you to contact. If you know that one mode or band is hosting some major event, you can almost always find a quiet spot on another mode or band.

The WARC bands (see the nearby sidebar "What's a WARC?") never have contests and usually are wide open for ragchews and casual operating.

Not-so-good times

Because a good ragchew lasts for a long time, pick a time and band that offer stable conditions. Propagation changes rapidly around sunrise and sunset. Local noon can be difficult on the higher bands. Don't be afraid to make contacts at any old time, though. You may surprise yourself and find out about propagation from the best teacher: experience.

Because weekends are busy times, you should check the contest and special-event calendars (see Chapter 11). A little warning keeps you from being surprised when you get on the air and allows you to be flexible in your operating.

When the bands seem to be frustratingly full, here are some helpful strategies that keep you doing your thing:

- **Try a nontraditional band.** Most nets and all contests are run on the traditional bands. The WARC bands almost always have sufficient space for a QSO.
- **Try a different mode.** Very few big contests have activity on more than one mode. You can change modes and enjoy a nice ragchew, too.

- **Be sure that you know how to operate your receiver.** Cut back on the RF gain, use narrower filter settings, know how to use controls such as the IF Shift and Passband Tuning controls, and generally be a sharp operator. You can remove much interference and noise just by using all the adjustments that your receiver provides.
- **Always have a backup plan.** There's no guarantee that any particular frequency will be clear on any given day. Hams have frequency freedom second to none, so use that big knob on the front of your radio.

Receivers are very sensitive and can easily be overloaded by strong signals, causing distortion that sounds like interference. You can make a huge improvement in listening quality by ensuring that your receiver is operating *linearly.* Start by turning off the preamp and noise blanker. Turn down the RF Gain control until the signal you're listening to is at the lowest comfortable level; you'll hear the noise background fade away. You can even switch on the receiver's attenuator to knock down strong signals even more. Use the minimum amount of sensitivity required to make the contact, and you'll enjoy listening a lot more.

Identifying a ragchewer

If you're in the mood for a ragchew, and you're tuning the bands, how can you tell whether a station wants to ragchew? The easiest way is to find an ongoing ragchew and join it. You can break in (see Chapter 8) or wait until one station is signing off and then call the remaining station.

Look for a station that has a solid signal — not necessarily a needle-pinning strong station, but one that's easy to copy and has steady signal strength. The best ragchews are contacts that last long enough for you to get past the opening pleasantries, so find a signal that you think will hold up.

One cue that the station isn't looking for a ragchew is a targeted call. You may tune in a PSK signal and see "CQ New York, CQ New York de W7VMI." W7VMI likely has some kind of errand or message and is interested in getting the job done. Perhaps the station on voice is calling "CQ DX" or "CQ mobiles." In that case, if you're not one of the target populations, keep on tuning.

Another not-a-ragchew cue is a hurried call or a call that has lots of stations responding. This station may be in a rare spot, in a contest, or at a special event. Keep tuning if you're really looking for a ragchew.

Calling CQ for a ragchew

Although responding to someone else's CQ is a good way to get started, it's also fun to go fishing — to call CQ and see what the bands bring.

The best CQ is one that's long enough to attract the attention of a station that's tuning by but not so long that that station loses interest and tunes away again. If the band is quiet, you may want to send long CQs; a busy band may require only a short CQ. As with fishing, try different lures until you get a feel for what works.

On voice modes, the key is in the tone of the CQ. Use a relaxed tone of voice and an easy tempo. Remember that the other station hears only your voice, so speak clearly, and be sure to use phonetics when signing your call. Sometimes, a little extra information — such as "from the Windy City" — helps attract attention. Don't overdo it, but don't be afraid to have a little fun.

On the digital modes, you don't want another station to have to watch line after line of characters scrolling off the screen, so keep your transmission reasonably short, like this example that shows a CQ message consisting of two lines. The "K" at the end of the second message is a *prosign* used for both Morse and digital modes to mean "I am done transmitting" when on voice you would say "Over":

CQ CQ CQ de NØAX NØAX NØAX CQ CQ CQ de NØAX NØAX NØAX K

When calling CQ on CW, I usually have good luck with 3-by-2-by-3 CQs (CQ CQ CQ de NØAX NØAX, repeated three times).

When you're sending a Morse code CQ (a general way to solicit a contact), don't send faster than you can receive. Having to ask the responding station to QRS (slow down) because you hustled through your CQ is embarrassing.

Evaluate on-the-air technique as you tune across the bands. Consider what you like and dislike about the various styles. Adopt the practices you like, and try to make them better; that's the amateur way.

Sharing a ragchew

Hams come from all walks of life and have all kinds of personalities, of course, so you'll come across both garrulous types, for whom a ragchew that doesn't last an hour is too short, and mike-shy hams, who consider more than a signal report to be a ragchew. Relax and enjoy the different people you meet.

If your radio has the capability to listen between the dits and dahs (which your operating manual calls the break-in or QSK feature), use it to listen for a station sending dits on your frequency. That means "I hear you, so stop CQing and let me call!" Then you can finish with "DE [your call] K," and the other station can call right away.

Roundtables — contacts among three or more hams on a single frequency — are also great ways to have a ragchew. Imagine getting together with your friends for lunch. If only one of you could talk at a time, that would be a roundtable. Roundtables aren't formal, like nets; they generally just go around the circle, with each station talking in turn. Stations can sign off and join in at any time.

Pounding Brass: Morse Code

CW is a lot easier to learn and copy if you're equipped to listen to it properly. For starters, headphones *(cans)* really help because they block out distracting noise. When you're copying code, your brain evaluates every little bit of sound your ears receive, so make its job easier by limiting noncode sounds.

Settling on one preferred pitch for the tones is natural, but over long periods, you can wear out your ear at that pitch. Keep the volume down and try different pitches so you don't fatigue your hearing.

When you have a comfortable audio environment, be sure that your radio is set up properly. Most radios come with a receiving filter intended for use with voice signals. Typically, a voice filter is 2.4 kHz wide (meaning that it passes a portion of radio spectrum or audio that spans 2.4 kHz) to pass the human voice clearly. CW doesn't need all that bandwidth. A filter 500 Hz wide is a better choice, and you can purchase one as an accessory for an older radio without adjustable software filters. The narrower filter removes nearby signals and noise that interfere with the desired signal. In fact, four or five code signals can happily coexist in the bandwidth occupied by a single voice signal.

Narrower isn't necessarily better below 400–500 Hz. A very narrow filter, such as 250 Hz models, may allow you to slice your radio's view of the spectrum very thin, but the tradeoff is an unnatural sound, and you'll be less able to hear what's going on around your frequency. These extra-narrow filters are useful when interference or noise is severe, but use a wider filter for regular operation.

Be sure to read the sections on CW operation in your radio's operating manual. Find out how to use all the filter adjustment controls, such as the IF Shift and Passband Tuning controls. Most CW operators like to set the AGC control to the FAST setting so that the radio receiver recovers rapidly. Being able to get the most out of your receiver is just as important on CW as on voice.

Copying the code

To get really comfortable with CW, you need to copy in your head. Watching a good operator having a conversation without writing down a word is an eye-opener. How does he do that? The answer is practice.

Getting started is always the hardest part, and a club of Morse enthusiasts is dedicated to helping you over that hurdle. The FISTS Club (`www.fists.org`) has training and practice resources; helps you on the air by providing a code buddy; and sponsors short, low-key (so to speak) operating events that are both fun and great practice.

As your code speed increases during the learning process, you gradually achieve the ability to process whole groups of characters as one group of sound. Copying in your head just takes that ability to another level. To get there, spend some time just listening to code on the air without writing anything down. Without the need to respond to the sender, you can relax and not get all tensed up, trying not to miss a character. Soon, you'll be able to hold more and more of the contact in your head without diminishing your copying ability.

When you try making Morse contact for real, jot down topics and information for your part of the next transmission on a piece of paper. (Resist the temptation to write each letter on the paper.)

Read some more mail on the bands, trying to relax as much as possible without staying right up with each character. Don't force the meaning; let your brain give it to you when it's ready. Gradually, the meaning pops into your head farther and farther behind the characters as they're actually received. What's happening is that your brain is doing its own form of error correcting, making sure that what you copy makes sense and taking cues from previous words and characters to fill in any blanks.

Good copying ability sneaks up on you over time. When you really hit a groove, you're barely conscious of the copying process at all. CW has become your second language.

Sending Morse

You may think that sending ability automatically follows receiving ability. To some extent, that's true, but after listening to other operators on the air, you'll find a wide range of sending ability. It isn't hard to develop a good, smooth sending style, or *fist*.

First, decide what type of device you want to use to send code. The basic telegraph key, or *straight key,* shown in Figure 9-7, is used on the bands every day, but sending good code at high speed with one is challenging. The straight key tops out at somewhere between 20 and 30 words per minute (wpm). At these speeds, sending becomes a full-body experience, and you have to be really skilled to make it sound good.

Combination paddle and key

Straight key

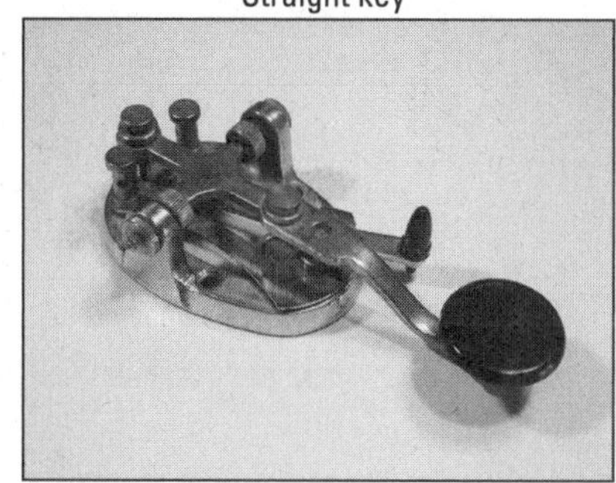

Semi-automatic key or "bug"

Figure 9-7: My paddle-key combo, a bug, and the venerable straight key.

You can find several better options. Before the advent of inexpensive electronics, fast code was sent with an *automatic key,* now known as a *bug,* shown in the middle of Figure 9-7. It's called a bug because the largest manufacturer, Vibroplex (`www.vibroplex.com`), uses a lightning bug as its symbol. ***Note:*** Bugs are rarely heard today, which makes their rhythm unusual and hard to copy, especially in the fist of an unskilled operator.

Electronic *keyers* are the most common way to send CW today. These devices generate dots and dashes electronically, controlled by the keying paddles. The simplest electronic keyers only make strings of dits and dahs; the operator puts them together with the right timing. More-sophisticated keyers make sure that the spacing between dits and dahs is correct.

More on Morse

Here are a few more tips for getting involved with Morse code:

- **Software programs:** You'll be a better Morse operator if you learn how to send and receive it yourself, but logging and ham-shack programs allow you to send Morse directly from your keyboard. A simple interface from a COM or USB port is all that's required to key your rig. Some of these programs include plug-ins that can copy code, too, as long as the frequency isn't too crowded or noisy. Visit `www.ac6v.com/morseprograms.htm` to find a Morse program to try.
- **Retro collections:** Collectors of Morse code equipment extend far beyond the ham radio community. Railroad and telegraph aficionados also have terrific collections of old keys, bugs, and paddles. For an entry into the world of antique code, start at the Sparks Telegraph Key Review (`www.zianet.com/sparks`). Also, Morse Express Books (`www.mtechnologies.com/books`) publishes excellent books about keys and related equipment.
- **New Year's Eve contacts:** Straight Key Night (`www.arrl.org/straight-key-night`) is a fun event that brings out old and new code equipment (and operators) around the world. Every New Year's Eve, hams break out their straight keys and bugs, and return to the airwaves for a few old-time QSOs before heading off to the evening's frivolities. An award is given for Best Fist, too. Give it a try this year.

The devices used to send code with keyers are called *paddles* (shown in Figure 9-7), referring to the flat ovals touched by the operator. A good operator can send well over 30 wpm with an electronic keyer and comfortable paddle. If you're serious about Morse, I recommend that you start with a paddle and keyer so that you don't have to change gears as your speed increases.

No matter whether you decide to start with a straight key or a paddle, use a good-quality instrument, for that's what it is: an instrument. See whether you can borrow or try one out. Experiment with different styles, and eventually you'll find one that feels just right. Key and paddle manufacturers include Vibroplex (mentioned earlier in this section), Bencher (`www.bencher.com`), and Begali (`www.i2rtf.com`).

Making Morse code contacts (CW)

Making code contacts, or CW, is a lot like making voice contacts in terms of structure. Hams are hams, after all. What's different about the Morse code contact is the heavy use of abbreviations, shorthand, and *prosigns* (two-letter combinations used to control the flow of a contact) to cut down the number of characters you send. You can find a complete list of CW abbreviations and prosigns at www.ac6v.com/morseaids.htm and in the *ARRL Operating Manual* (www.arrl.org/shop/operating).

After you begin a contact and exchange call signs, giving your call sign every time you turn the transmission over to the other station isn't necessary, but you must include it once every ten minutes, as required by FCC rules. Send your information and end with the "BK" prosign to signal the other station that he or she can go ahead. This method is much more efficient than sending call signs every time.

While you're listening to Morse, you may hear some odd characters that don't make up a word or abbreviation, especially as the operators start and stop sending. These characters are *prosigns* (short for *procedural signs)* used to control who sends and who doesn't. On voice, "over" is the same as the prosign K. Some prosigns, such as BK (break), are two letters sent together. Others include SK, which means "end of contact," and KN, which means "only the station I am in contact with start transmitting." You can find out more about prosigns and other Morse conventions and abbreviations at www.hamuniverse.com/qsignals.html.

At the conclusion of a Morse code contact, after all the 73s (best regards) and CULs (see you later), be sure to close with the appropriate prosign: SK for "end of contact" or CL if you're going off the air. You may also hear the other station send "shave and a haircut" (dit-dididit-dit), and you're expected to respond with "two bits" (dit dit). These rhythms are deeply ingrained in ham radio and even occur in spoken conversations between hams. I wrap up many a chat with "diddly bump-de-bump," the rhythm of SK, or just "dit dit," meaning "See ya!" Yeah, it's a little goofy, but have fun!

Receiving Messages Afloat and Remote

Once upon a time, any kind of messaging capability was tied to specific phone numbers, bulletin boards, and servers. If you weren't in range of your home bulletin board or service provider, you were pretty much disconnected. Ham radio has made those days obsolete, just as wireless networks have freed the computer from cabled connections.

Hams can use a computer and a radio to connect directly to gateway stations around the world via terrestrial links on the HF bands or via VHF and UHF to the amateur satellites. The gateway stations transfer e-mail messages between ham radio and the Internet. The satellites transfer messages to and from the Internet via a ground control station. Either way, a ham far from home can use ham radio to send and receive e-mail.

Using Winlink

The best-known HF message system is Winlink 2000, referred to as Winlink or WL2K, which enables any ham to send and receive e-mail by using the PACTOR or WINMOR digital mode. (I talk about digital modes in Chapter 11.) Winlink (see Figure 9-8) is a worldwide network of stations operating 24 hours a day on the HF bands as well as on VHF and UHF via packet radio. It has grown from a network used by boaters to a sophisticated, hardened network used for disaster relief and emergencies of all kinds.

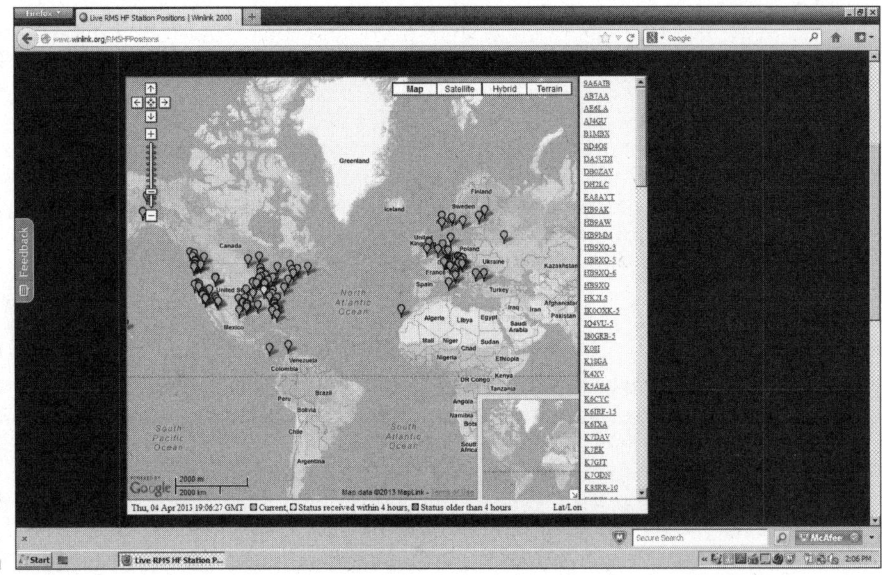

Figure 9-8: Winlink digital data network stations as of March 2013.

To find the frequencies of Winlink stations, visit `www.winlink.org/stations`. This extensive and growing network covers much of the Earth. These stations are linked via the Internet, creating a global home for Winlink users. The Winlink system is also connected to the Automatic Packet Reporting System (APRS; `www.aprs.org`) so that position, weather, and other information can be exchanged or viewed via the Internet. (I discuss APRS in Chapter 11.)

To use the Winlink system, you must register as a user on the Winlink network so that the system recognizes you when you connect. When you're a recognized user, your messages are available from anywhere on Earth, via whichever Winlink station you use to connect. You must also download and install a Winlink-compatible e-mail program, such as Airmail, which is available on the Winlink website.

Along with a computer that runs the e-mail client software, such as the popular Airmail (www.siriuscyber.net/ham), you need a way to generate the signals for the digital mode you choose to access the Winlink Radio Mail Server (RMS) stations. On HF, you need a sound card and software to send and receive via WINMOR mode (www.winlink.org/WINMOR) or an external communications processor that supports the PACTOR family of digital modes, such as the SCS PTC or P4 modem (www.scs-ptc.com/pactor). On VHF and UHF, you can use a standard packet-radio terminal node controller to connect with a local RMS station.

Connecting with Airmail and Winlink

If you're using Airmail and are registered with the Winlink network, the connection process is straightforward. Follow these steps to connect:

1. **Open the Airmail program.**
2. **Click the Terminal icon.**

   A menu of Winlink stations and frequencies pops up.
3. **Select a station and frequency appropriate for your location, time of day, and equipment.**

   The computer and radio attempt to connect to the Winlink server and notify you of success or failure. This process is very much like a Wi-Fi card connecting to the Internet through a wireless data router at a library or coffee shop.

The usual regulations on ham radio messages apply, of course. You can't encrypt messages, send business traffic or obscene content, or use radio links on behalf of third parties in countries where such use is prohibited. Winlink isn't a web browsing service, but web content is often added to e-mail. That content isn't available via the Winlink system.

Don't forget that all the usual operating protocols and requirements still apply. You have to listen to the channel and make sure you won't interfere with ongoing communications, and you have to monitor your transmissions so that you fulfill your obligations as a control operator.

Also keep in mind that although the Winlink stations are connected via the Internet, your station is connected by a relatively slow digital data radio link. The data rate is limited due to the radio link, so don't try to send big files or messages. Nevertheless, Winlink 2000 is a tremendous service and a boon to those who travel off the beaten track and to those who need to communicate during and after emergencies.

Chapter 10

Operating with Intent

In This Chapter

- Registering with an emergency organization
- Getting ready for and operating in emergencies
- Rendering public service
- Transmitting on nets
- Knowing how to handle traffic

As your experience with ham radio grows, you'll find more and more practical uses for your communications skills. Your ham radio skills can also benefit others, which is where the *service* part of *amateur radio service* comes in.

In return for the privileges that go with the license — access to a broad range of frequencies, protection from many forms of interference, maintenance of technical standards, and enforcement of operating rules — amateurs give back by providing emergency communications systems and trained operators. Between emergencies, hams also provide communications support at public functions and sporting events, keep an eye on the weather, and perform training exercises.

These services are important to you for two reasons: You can use them, and you can provide them. In this chapter, you find out what those services are and how to get started with the groups that provide them.

Note: This chapter, written primarily for American and Canadian hams, describes the U.S. emergency communications organizations. Elsewhere around the world, you can find similar organizations. Contact your national amateur radio society for information about them.

Joining an Emergency Organization

An important element of the Federal Communications Commission's (FCC's) Basis and Purpose of amateur radio is emergency communications. Just during the few months that I spent writing this book, hams provided critical emergency communications during hurricanes on the East Coast, wildfires across the West, and tornadoes and floods in the Midwest. You never know when an emergency will arise, so start preparing yourself as soon as you're licensed.

Emergency communications (known in the radio biz as *emcomm)* is loosely defined as any communication with the purpose of reducing an immediate threat of injury or property damage — everything from reporting car accidents to supporting large-scale disaster relief. In this section, I introduce the elements of emcomm and show you how to get started.

Finding an emcomm group

Whether your interest in emcomm is to support yourself and your family or to participate in organized emcomm, you need to know how amateurs are organized. Otherwise, how will you know where to tune or how to interact with them?

This section gives you several good places to start.

The different levels of emergencies and disasters, with varying degrees of resource requirements, require different responses by government agencies. As a result, a single, one-size-fits-all amateur emergency organization isn't enough to handle all emergencies.

ARES

ARRL's Amateur Radio Emergency Service (ARES) is the largest nationwide ham radio emcomm organization, organized by individual ARRL sections that may be as large as a state or as small as a few counties, depending on population. ARES is managed by the ARRL Field Organization (a system of volunteer managers and technical resources) and works primarily with local public safety groups and nongovernmental agencies, such as local fire departments and the American Red Cross. Local ARES leaders determine how best to organize the volunteers and interact with the agencies their groups serve. Training is arranged by the ARES teams and local organizations.

For complete information about ARES, check out the online Public Service Communications Manual at `www.arrl.org/public-service-communications-manual`.

RACES

Radio Amateur Civil Emergency Service (RACES), organized and managed by the Federal Emergency Management Agency (FEMA), is a national emergency communications organization governed by special FCC rules. Its mission, like that of ARES, is to provide communications assistance to public and private agencies during a civil emergency or disaster. The organization is open to all amateurs and welcomes your participation.

RACES groups are organized and managed by local, county, or state civil-defense agencies that are responsible for disaster services and activated during civil emergencies by state or federal officials. RACES members are required to be members of their local civil-preparedness groups as well, and they receive training to support those groups.

MARS

A third organization that maintains an extensive emergency communications network of ham volunteers is the Military Auxiliary Radio System (MARS), which provides an interface between the worldwide military communications systems and ham radio. MARS is sponsored by the U.S. Department of Defense, and each branch of the military has its own MARS program.

MARS members receive special licenses and call signs that allow them to operate on certain MARS frequencies just outside the ham bands. MARS provides technical and operations training, as well as preparation for emergency communications.

Preparing for national emergencies

Over the past decade, emergency response organizations at all levels of management have come together to develop a common plan for responding to emergencies and disasters. This plan — the National Incident Management System (NIMS) — is mandatory for all public-safety and other government agencies in the United States. Having a common plan greatly improves the capability of different organizations to work together efficiently and effectively.

Because these agencies are primary customers for amateur radio emergency communications, ARES teams have adopted NIMS and made it a central part of their training, as have many other nongovernmental organizations. To find out more about NIMS, download the NIMS overview document at `www.fema.gov/national-preparedness/national-incident-management-system`. Your emergency response team will show you more as it applies to your community.

FEMA also provides free training courses at `http://training.fema.gov/IS/NIMS.aspx`. Your ARES or other emcomm team will help you get the right training and certification for your area.

Volunteering your services

You can volunteer for ARES, RACES, and MARS as follows:

- **ARES:** You can register as an ARES volunteer simply by filling out the form at www.arrl.org/files/file/Public%20Service/fsd98.pdf and mailing it to the ARRL. You also need to join a local ARES team to participate in training and exercises. The easiest way to find out about the ARES organization in your area is to contact your ARRL section manager, listed at www.arrl.org/sections. You can also search www.arrl.org/arrl-net-directory-search for ARES nets in your area; check in to the net as a visitor, and ask for information about ARES.
- **RACES:** You can get more information about RACES by contacting the civil-defense organization in your area, which is managed by your county or parish (or the local equivalent).
- **MARS:** To be a MARS volunteer, you must be at least 18 years old, be a U.S. citizen, and hold a valid amateur license. For more information on the MARS program, check out the Army's MARS Facebook page at https://www.facebook.com/HQArmyMARS.

I recommend that you start by participating in ARES. Then, if you like being an ARES member, dual membership in ARES and RACES may be for you. The ARES-RACES FAQ page at www.arrl.org/ares-races-faq helps explain how the two services relate to each other.

If you want to help administer and manage emcomm activities in your ARRL section after you have some ham radio experience, consider applying for an ARRL Field Organization appointment. Beginning emcomm volunteers like you can fill the following positions:

- **Assistant section manager (ASM):** The section manager (SM) is appointed, but you can always assist him or her. Tasks vary according to the activities of the section, but typical duties include collecting and analyzing volunteer reports, and working with and checking into local and regional nets. Should a special task arise, you may be asked to perform it on behalf of the SM.
- **Official emergency station (OES):** As the control operator of an OES, you perform specific actions as required by local emergency coordinators. OES appointments go to stations that are committed to emcomm work; they provide the opportunity to tackle detailed projects in operations, administration, or logistics.
- **Public information officer (PIO):** You can establish relationships with local and regional media to publicize ham radio, particularly the public

service and emcomm performed on behalf of the public. PIOs also help establish good relationships with community leaders and organizations.

- **Official observer (OO):** OOs help other hams avoid receiving an FCC notice of rule violation because of operating or technical irregularities. They also keep an ear out for unlicensed intruders or spurious transmissions from other services.
- **Technical specialist (TS):** If you have expertise in a specific area, or if you're generally skilled in some aspect of radio operations, you can be a technical specialist. A TS serves as a consultant to local and regional hams, as well as to the ARRL.

Check out other interesting section- and division-level appointments at `www.arrl.org/field-organization`.

Preparing for an Emergency

Getting acquainted with emergency organizations is fine, but it's only a start. You need to take the necessary steps to prepare yourself so that when the time comes, you're ready to contribute. Preparation means making sure that you know four things:

- Who to work with
- Where to find emcomms groups on the air
- What gear to have on hand
- How to be of service

Knowing who

Earlier in this chapter, I discuss the organizations that provide emergency communications. First, become familiar with the leaders in your ARRL section; then get acquainted with the local team leaders and members.

The call signs of the local clubs and stations operating from governmental emergency operations centers (EOC) are valuable to have at your fingertips in times of emergencies. The best way to get familiar with these call signs (and make your call sign familiar to them) is to be a regular participant in nets and exercises. Checking in to weekly nets takes little time and reinforces your awareness of who else in your area is participating. If you have the time, attending meetings and other functions such as EOC open houses and work parties also helps members put faces with the call signs. Building personal relationships pays off when a real emergency comes along.

Knowing where

When an emergency occurs, you don't want to be left tuning around the bands trying to find the emcomm groups. Keep a detailed list of the emergency net frequencies, along with the names of the leaders in your area. (I provide a link to a downloadable chart for you to fill in at www.dummies.com/cheatsheet/hamradio.) You may want to reduce this list with a photocopier and laminate it for a long-lasting reference the size of a credit card that you can carry in your wallet or purse.

Knowing what

If an emergency occurs and your equipment isn't ready, you can be under tremendous pressure. In your haste, you might omit some crucial item or won't be able to find it on the spur of the moment. I recommend assembling a go kit (similar to a first-aid kit) as an antidote to this adrenaline-induced confusion.

Assembling a go kit

A *go kit* is a group of items that will be necessary during an emergency; you collect these items in advance and place them in a handy carrying case. Then, if an emergency situation actually arises, the go kit allows you to spend your time responding to the emergency instead of racing to get your gear together. Preparing the kit in advance also makes you less likely to forget important elements.

Figure 10-1 shows a portable go kit that can sustain a couple of people for 24 hours.

Before making up your go kit, consider what mission(s) you may be attempting. A personal checklist is a good starting point. You can find a generic checklist in the ARES Field Resources Manual at www.arrl.org/files/file/ARESFieldResourcesManual.pdf.

What goes into a go kit varies from ham to ham, but every kit should contain the following essentials:

- **Nonperishable food:** During an emergency, you won't know when your next meal will arrive. Remove the uncertainty by having your own food (the kind that doesn't require refrigeration). If you bring canned food, don't forget the can opener!

- **Appropriate clothing:** If you get too cold, you'll want a jacket nearby; if you get too hot, you'll want to exchange your current clothing for something more lightweight. Preparation allows flexibility.
- **Radios and equipment:** Don't forget to bring all the equipment you may need: radios, antennas, and power supplies. Make sure everything is lightweight, flexible, and easy to set up.
- **References:** You need lists of operating frequencies, as well as phone lists — a personal phone list and a list of emergency-related telephone numbers.

For a complete list of go kits for every occasion and need, check out Dan O'Connor's Personal Go Kit for Emergency Communications presentation at `www.ke7hlr.com/ecw/personal_go-kit_2011.pdf`.

Preparing your home

You may not need a formal go kit if you operate from home, but you still need to prepare for emergencies such as an extended power outage or the failure of your main antenna.

Figure 10-1: A go kit.

Courtesy Ralph Javins (N7KGA)

Your primary concern is emergency power. Most modern radios aren't very battery friendly, drawing more than 1 amp even when they're just receiving. You'll need a generator to power them during any extended power outage. If you have a home generator, make sure that you can connect to it safely and that it can adequately power the AC circuits in your radio shack.

If you don't have a generator, you may be able to use another backup power source: Most radios with a DC power supply can run from an automobile battery. Getting power from your car to your radio isn't always easy, however. Decide which radios you want to operate from your car, and investigate how you can power and connect an antenna to each of them.

Overall, the most important step is to simply attempt to answer this question: "How would I get on the air if I'm unable to use my regular shack?" Just by thinking things through and making plans, you'll be well on the road to being prepared for any emergency.

Knowing how

Knowing the procedures to follow is the most important part of personal preparedness. Whatever your experience and background are, you have to know the specific details of working with your emergency organizations. If you don't, you won't be prepared to contribute when you show up on the air from home or at a disaster site.

Do everyone a favor — including yourself — by spending a little time getting trained in the necessary procedures and techniques. Your local emcomm organization has plenty of training opportunities and training nets for practice. Participating in public-service activities, such as acting as a race-course checkpoint in a fun run or as a parade coordinator, is awfully good practice, and it exercises your radio equipment as well. (By the way, you'll make good friends at these exercises who can teach you a lot.)

The ARRL offers emergency-communications training courses that you can take online. Check out these courses at `www.arrl.org/emergency-communications-training`.

After you start training in emergency communications, you'll find that training is available for many other useful skills, such as cardiopulmonary resuscitation (CPR), first aid, orienteering, and search and rescue.

Operating in an Emergency

All emergencies are different, of course, so no single step-by-step procedure is always going to be useful. But here are some solid general principles, based on the ARES Field Resources Manual, to follow when disaster strikes:

1. **Make sure that you, your family, and your property are safe and secure before you respond as an emcomm volunteer.**
2. **Monitor your primary emergency frequencies.**
3. **Follow the instructions you receive from the net control or other emergency official on the frequency.**

   Check in if and when check-ins are requested.
4. **Contact your local emergency communications leader or designee for further instructions.**

Everyone is likely to be fairly excited and tense. Keep your head on straight and follow your training so that you can help rather than hinder in an emergency situation.

Reporting an accident or other incident

Accident reports are more common than you may think. Anybody who spends time driving can attest to the frequency of accidents, and many rural areas have little or no mobile phone service. I've personally used ham radio to report accidents, stalled cars, and fires. Know how to report an incident quickly and clearly — and don't assume that people with mobile phones are already doing it.

Follow these steps to report an incident via a ham repeater's autopatch:

1. **Turn up your radio's output power to maximum, and clearly say "Break" or "Break emergency" at the first opportunity.**

   A strong signal can get the attention of listening stations. Don't shy away from interrupting an ongoing conversation.
2. **After you have control of the repeater or the frequency is clear, state that you have an emergency to report.**
3. **State clearly that you're making an emergency autopatch and then activate the autopatch system.**

   See Chapter 9 for details on repeater autopatch capabilities.

If you can't activate the repeater's autopatch, ask another repeater user to activate it for you, or ask that someone make an emergency relay to 911. Report all the necessary material and then stand by on frequency until the relaying station reports to you that the information has been relayed and the call is complete.

4. **Dial 911, and when the operator responds, state your name and say that you're reporting an emergency via amateur radio.**

5. **Follow the directions of the 911 operator.**

   If the operator asks you to stay on the line, do so, and ask the other repeater users to please stand by.

6. **When the operator finishes, release the autopatch, and announce that you've released it.**

Whether you use a repeater's autopatch feature or relay the report via another amateur, you need to be able to generate clear, concise information. To report an automobile accident, for example, you should know the following details:

- The street name or highway number
- The street address or approximate highway mile marker
- The direction or lanes in which the accident occurred
- Whether the accident is blocking traffic
- Whether injuries are apparent
- Whether the vehicles are on fire, are smoking, or have spilled fuel

Don't guess if you don't know something for sure! Report what you know, but don't embellish the facts.

Making and responding to distress calls

Before an emergency occurs, be sure that you know how to make a distress call on a frequency where hams are likely to be listening, such as a marine service net or a wide-coverage repeater frequency. Store at least one of these frequencies in your radio's memory, if possible.

Anyone, licensed or not, can use your radio equipment in an emergency to call for help on any frequency. You won't have time to be looking at net directories in an emergency.

Making a distress call

Do the following things when you make a distress call:

1. **For immediate emergency assistance, say "Mayday" or send the Morse code signal SOS (yes, just like in the movies).**

   Maydays sound something like "Mayday, mayday, mayday, this is NØAX," followed by

   - The location (latitude/longitude) or address of the emergency
   - The nature of the emergency
   - The type of assistance needed (such as medical or transportation aid)

2. **Repeat your distress signal and your call sign for several minutes or until you get an answer.**

   Even if you don't hear an answer, someone may hear you.

3. **Try different frequencies if you don't get an answer.**

   If you decide to change frequencies, announce the frequency to which you're moving so that anyone who hears you can follow.

Responding to a distress call

Here's what to do if you hear a distress signal on the air:

1. **Immediately record the time and frequency of the call.**
2. **Respond to the call.**

   Say something like this: "[The station's call sign], this is [your call sign]. I hear your distress call. What is your situation?"

3. **Collect and record the following information:**

   - The location (latitude/longitude) or address of the emergency
   - The nature of the problem
   - The type of assistance needed (such as medical or transportation aid)
   - Any other information that might help emergency responders

4. **Ask the station in distress to remain on frequency.**
5. **Call the appropriate public agency or public emergency number, such as 911.**

   Explain that you're an amateur radio operator and that you've received a distress call. The dispatcher will either ask you for information or transfer you to a more appropriate agency.

6. **Follow the dispatcher's instructions to the letter.**

   The dispatcher may ask you to act as a relay to the station in distress.

7. **As soon as possible, report back to the station in distress.**

   Tell the operator whom you contacted and any information you've been asked to relay.

8. **Stay on frequency as long as the station in distress or the authorities need your assistance.**

Supporting emergency communications outside your area

How can you provide assistance in case of a disaster or emergency situation outside your immediate vicinity? The best thing you can do is make yourself available to the on-site communications workers — but only if you're called upon to do so. Because most of the important information from a disaster flows out, not in, you don't want to get in the way.

If a hurricane is bearing down on Miami, for example, getting on the air and calling "CQ Miami!" is foolish. You have only a minimal chance of actually rendering assistance, but you stand a good chance of misdirecting some actual emergency communication by the proper authorities. Instead, support the communications networks that the Miami hams depend on. Check in to your NTS local nets to see whether any messages need to be relayed to your location. Monitor the Hurricane Watch Net on 14.325 MHz and any Florida emergency net frequencies. Tune to the bands that support propagation to Florida, in case someone is calling for help; you may be able to relay information from a station that's unable to contact local authorities.

Here's another example. Suppose that a search-and-rescue (SAR) operation in the nearby foothills is taking place with nets on 2 meter repeaters and several simplex frequencies coordinating the activities. Do you check in to the SAR nets? No! But you can *monitor* (listen without transmitting) their operations to see whether an opportunity arises for you to provide assistance, especially if you have beam antennas that you can aim directly at the area. (See Chapter 12 for more on beam antennas.) If you can set your radios to listen to the repeater input frequencies as well as the outputs, you may hear a weak station that's unable to activate the repeater. If you monitor the simplex frequencies, you may act as a relay station. Two stations in hilly areas may be unable to communicate directly, but you can hear both and can relay communications between them. If such a situation occurs, you can break in and say, "This is [your call sign], and I can copy both stations. Do you want me to relay?"

You need to help information flow out from the disaster site, not force more in. Listen, listen, listen. That's good advice most of the time.

Providing Public Service

Between emergencies, hams perform many other valuable public services. After you become associated with a local emergency communications group, you can use your ham radio skills for the public's benefit.

Weather monitoring

One of the most widespread public-service functions is amateur weather watching. In many areas, particularly those that have frequent severe weather conditions, nets devoted to reporting local weather conditions meet regularly. Some nets meet once or twice every day; others meet only when a threat of severe weather exists.

Many weather nets are associated with the Skywarn program (`www.skywarn.org`), operated by the National Oceanic and Atmospheric Association (NOAA). Groups reporting weather conditions under the Skywarn program relay information to the National Weather Service (NWS), which uses the reports in forecasting and severe weather management. In some areas, a net control station may operate a station from the NWS itself. For information on whether a Skywarn net is active in your area, click the Local Groups link on the Skywarn home page or enter *Skywarn net* in an Internet search engine.

Other weather nets may operate on VHF/UHF repeaters or on 75 meter voice nets. The New England Weather Net, for example, meets on 3905 kHz every day at 5:30 a.m. Informal weather nets on local repeaters are common. Ask around to see if one operates in your area. These nets usually are active at commuting drive times. (For more information on these groups, see "Participating in Nets," later in this chapter.)

Parades and sporting events

Amateurs participate in parades and sporting events to give the event managers timely information and help with coordination. In return, amateurs get good training in communications procedures and operations that simulate real-life emergencies. You can think of a parade, for example, as being similar to a slow-speed evacuation. A lost-child booth at a parade is similar to a small SAR operation. Helping keep track of race entrants in a marathon or bikeathon is good practice for handling health-and-welfare messages.

A leader of the amateur group usually coordinates plans with the event managers; then the group deploys as the plan requires. Depending on the size of the event, all communications may take place on one frequency, or several channels may be required. Information may be restricted to simple status, or actual logistics information may be relayed. Communications support includes a wide variety of needs.

Event managers typically work with a single club or emergency-services group that manages the ham radio side of things. If you want to participate in these events, start by contacting your section manager. He or she can direct you to one or more people active in public service who can let you know about upcoming events.

Here are a few tips for supporting an event:

- Get the appropriate identification and any required insignia, and dress similarly to the rest of the group members.
- Take along a copy of your amateur license and a photo ID.
- Take water and some food with you in case you're stationed somewhere without support.
- Don't assume that you'll be out of the weather. Protect yourself against the elements.
- Have your identification permanently engraved or attached to your radio equipment to protect against theft.

Remember the prohibition against hams being paid for communicating on behalf of third parties! You and your emcomm team may be reimbursed for direct expenses (such as mileage or materials) but not compensated in any other way. It's up to the leaders of your team to ensure that your support of any event or organization complies with this important rule.

Participating in Nets

Nets — regularly scheduled on-the-air meetings of hams who have common interests — are among the oldest ham radio activities. The first net was probably formed as soon as two hams went on the air. Sometimes, the nets are strictly for pleasure, to discuss topics such as collecting things, playing radio chess, or pursuing awards. Other nets are more utilitarian, such as those for traffic handling, emergency services, and weather reporting.

If a net follows standard operating rules, it's called a *directed net.* Nearly all directed nets have a similar basic structure. A net control station (NCS) initiates the net operations, maintains order, directs the net activities, and then terminates net operations in an orderly way. Stations that want to participate in the net check in at the direction of the NCS. A net manager defines net policy and focus, and works with the NCS stations to keep the net meeting on a regular basis.

Nets are run in many ways. Some nets are formal; others are more like extended roundtable QSOs. The key is to listen, identify the NCS, and follow the directions. The behavior of other net members is your guide.

Checking in

If you want to check in to a net, you register your call sign and location or status with the NCS. Be sure that you can hear the NCS clearly and that you can understand his or her instructions. If you're not a regular net member, wait until the NCS calls for visitors. When you check in, give your call sign once (phonetically if you're using voice). If the NCS doesn't copy your call sign the first time, repeat your call sign, or the NCS can ask one of the listening stations to relay your call sign.

You can check in with business (such as an announcement) or traffic (messages) for the net in a couple of ways; listen to the net to find out which method is appropriate. The most common method is to say something like "NØAX with one item for the net." The NCS acknowledges your item, and you wait for further instructions. Alternatively, you can check in with your call sign, and when the NCS acknowledges you and asks whether you have any business for the net, reply, "One item." Listen to other net members checking in, and when in Rome, check in as the Romans do.

If you want to contact one of the other stations checking in, you can declare this intention when you're checking in as though it were net business or wait until the check-in process is complete and the NCS calls for net business. Either way, the NCS asks the other station to acknowledge you and puts the two of you together, following net procedures.

Exchanging information

Information such as formal radiograms (see Figure 10-2) or verbal messages are exchanged during a net, either on frequency or off frequency, while the net operations continue. Nets that exist primarily for discussion of a common interest or for selling and trading equipment, for example, tend to keep all their transmissions on one frequency so that everyone can hear them. This system is quite inefficient for a net that's intended to route messages, such as in an emergency, so the NCS sends stations off frequency to exchange the information and then return to the net frequency.

THE AMERICAN RADIO RELAY LEAGUE
RADIOGRAM
VIA AMATEUR RADIO

NUMBER	PRECEDENCE	HX	STATION OF ORIGIN	CHECK	PLACE OF ORIGIN	TIME FILED	DATE
42	R	HXG	NØAX	14	VASHON, WA	4:15AM	2 Oct 03

TO CONTEST MANAGER
ORANGE JUICE CONTESTERS
BIG SQUEEZE, FLORIDA
555-555-1212

TELEPHONE NUMBER

THIS RADIO MESSAGE WAS RECEIVED AT
AMATEUR STATION PHONE
NAME
STREET ADDRESS
CITY AND STATE

HAD	FUN	IN	THE	OJ
CONTEST	YESTERDAY	X	WHEN	IS
THE	NEXT	ONE	?	

73, WARD NØAX

REC'D FROM DATE TIME

SENT TO W7KT DATE 3 Oct 03 TIME 12:05 PM

THIS MESSAGE WAS HANDLED FREE OF CHARGE BY A LICENSED AMATEUR RADIO OPERATOR, WHOSE ADDRESS IS SHOWN IN THE BOX AT RIGHT ABOVE. AS SUCH MESSAGES ARE HANDLED SOLELY FOR THE PLEASURE OF OPERATING, NO COMPENSATION CAN BE ACCEPTED BY A "HAM" OPERATOR. A RETURN MESSAGE MAY BE FILED WITH THE "HAM" DELIVERING THIS MESSAGE TO YOU. FURTHER INFORMATION ON AMATEUR RADIO MAY BE OBTAINED FROM A.R.R.L. HEADQUARTERS, 225 MAIN STREET, NEWINGTON, CONN 06111

THE AMERICAN RADIO RELAY LEAGUE, INC., IS THE NATIONAL MEMBERSHIP SOCIETY OF LICENSED RADIO AMATEURS AND THE PUBLISHER OF QST MAGAZINE. ONE OF ITS FUNCTIONS IS PROMOTION OF PUBLIC SERVICE COMMUNICATIONS AMONG AMATEUR OPERATORS. TO THAT END, THE LEAGUE HAS ORGANIZED THE NATIONAL TRAFFIC SYSTEM FOR DAILY NATION-WIDE MESSAGE HANDLING

PRINTED IN U.S.A.

Figure 10-2: An ARRL radiogram form with a sample message.

Here's an example of an NCS directing an off-frequency message exchange during an emergency communications net that's using a repeater. In this exchange, *traffic* means a single message, and *EOC* stands for *emergency operations center.*

W2—: I have one piece of traffic for the EOC.

(W2— is either relaying the message from another ham or serving as the originating station.)

NCS: W2—, stand by. EOC, can you accept traffic?

(The net is practicing the use of tactical call signs along with the FCC-issued call signs.)

EOC: EOC is ready for traffic.

NCS: W2— and EOC, move to the primary simplex frequency and pass the traffic.

This transmission means that W2— and the station at the EOC are to leave the net frequency, change to the team's primary simplex (no repeater) frequency, and reestablish contact, after which W2— transmits the message to the EOC. When both stations are done, they return to the net frequency and report to the NCS.

As a control operator, you still have to give your FCC call sign along with the tactical call sign whenever you begin operation, every ten minutes during operation, and whenever the operator changes. Your emcomm team will provide training to help you satisfy these simple rules.

Handling Traffic

The National Traffic System (NTS; www.arrl.org/nts), managed by the ARRL Field Organization, is the backbone of all amateur message handling in North America. The NTS runs every day, during emergencies and normal times alike. Traffic handling is less popular an activity than it was in years past, but it builds skilled and accurate operating techniques, making you ready for any emergency.

Why, in this age of wireless networks and coast-to-coast links, does traffic handling still exist? The main reason is that when all else fails — and sophisticated communications systems such as mobile phones and the Internet can fail quickly during a disaster or emergency — ham radio traffic handlers fill the gap in an accurate, accountable way until those faster systems are brought back online. Under such circumstances, large numbers of simple text messages must be sent quickly and reliably. Health-and-welfare messages stream out of the afflicted areas until more-sophisticated systems are brought in or restored. Even when digital modes are used to send the message, radiograms (refer to Figure 10-2, earlier in this chapter) are employed to make the process accountable and traceable, as good emergency management requires.

Traffic handling is supposed to work in an emergency, when the chips are down and only the barest minimum of resources may exist.

The *ARRL Operating Manual* (www.arrl.org/shop/operating) has an extensive chapter on handling traffic procedures and jargon.

Chapter 11

Operating Specialties

In This Chapter

- Contacting distant stations
- Participating in contests
- Pursuing awards
- Operating with low power
- Mastering Morse code
- Exploring the digital modes
- Using satellites
- Communicating with images

After you get rolling with casual operating, you can begin to explore a whole world of interesting specialties within ham radio. Specialties, to many, are the real attraction of the hobby.

In this chapter, I give you an overview of the most popular activities, cover some of the basic techniques and resources, and demystify a little bit of the specialized jargon of ham radio. I don't hit all the possible activities by any means, but start with the activities I discuss in this chapter, and you'll discover many others along the way, especially if you read the magazines and browse the popular websites.

DXing

Pushing your station to make contacts over greater and greater distances, or *DXing* (*DX* means *distance*), is a driving force that fuels the ham radio spirit. Somewhere out in the ether, a station is always just tantalizingly out of reach; the challenge of contacting that station is the purpose of DXing.

Thousands of hams around the world like nothing better than making contacts (QSOs) with people far away. These hams seem to ignore all nearby stations, because their logs are filled with exotic locations. Ask them about some odd bit of geography, and you're likely to find that they not only know where it is, but also know some of its political history and the call sign of at least one ham operator there. These hams are *DXers.*

The history of ham radio is tightly coupled with DXing. As transmitters became more powerful and receivers more sensitive, the distances over which a station could make contact were a direct measure of its quality. Hams quickly explored the different bands and followed the fluctuations of the ionosphere. DXing drove improvements in many types of equipment, too, especially antennas and receivers.

Today, intercontinental contacts on the HF frequencies traditionally considered to be the shortwave bands are common but still thrilling. Cross-continental contacts on VHF and UHF, once thought impossible, are being made in increasing numbers. Because the Sun and the seasons are always changing, each day you spend DXing is a little (and sometimes a lot) different. Sure, you can log in to an Internet chat room or send e-mail around the world, but logging a QSO, mastering the vagaries of the ionosphere, and getting through to a distant station are real accomplishments.

Listen for *DXpeditions* — special trips made to remote or unusual locations by one or more hams just for the purpose of putting them on the air.

DXing on the shortwave bands

The following sections show you how to use the shortwave or HF bands to contact a distant station, a process known as *working DX.* These bands are at frequencies below 30 MHz on which signals routinely travel around the Earth, bouncing between the ionosphere and the planet's surface as they go. Signals can be exchanged over long distances on the shortwave bands easily, so HF DXing is often a worldwide event, with stations calling from several continents.

VHF and UHF DXing are no less exciting on these bands but require a different approach. Propagation is much more selective here, so fewer signals are detectable at one time and usually come from stations concentrated in a few areas. These differences make DXing quite different on the shortwave bands versus VHF and UHF — a topic that I cover in "DXing on the VHF and UHF bands," later in this chapter.

If you want more information about propagation, check out the *Ham Radio For Dummies* website (www.dummies.com/extras/hamradio). For in-depth information about shortwave DXing techniques, I recommend *The Complete DXer,* by Bob Locher (W9KNI), published by Idiom Press. Now in its third edition, Locher's book has Elmered legions of beginning DXers.

Picking up DX signals

Even if you have a very modest home or mobile HF station, you can work DX. Skill and knowledge compensate for a great deal of disparity in equipment. The first skill to master is how to listen.

Signals coming from far away have to make several hops off the ionosphere and take several paths to your station. The signals mix with one another as they arrive at your antenna, spreading out in time and changing strength rapidly. You need to be able to recognize accents and signals that have a curious, hollow, or fluttery sound, because they mean that DX is at hand.

Subscribe to one of the online DX newsletters, such as The Daily DX (www.dailydx.com) and The QRZ DX Weekly Newsletter (www.dxpub.com/qrz_dx_nl.html). Watch these publications for upcoming or current DX station activity, and program the expected operating frequencies into your rig's memory for easy access.

Start tuning from the bottom of the band, and keep listening, noting what you hear and at what times. In DXing, experience with the characteristics of a band's propagation is the best teacher. Try to detect a pattern when signals from different population centers appear, and see how the seasons affect propagation on the different bands. Soon, you'll recognize the signals of regulars on the band, too.

If you plan on doing a lot of DXing, bookmark the ARRL's DXCC Award program website at www.arrl.org/dxcc, and purchase or download a copy of the ARRL DXCC List (www.arrl.org/files/file/DXCC/2013%20DXCC%20Current_a.pdf) and a ham radio prefix map of the world for reference. Maps are often available for free from manufacturers like Yaesu and Icom, or you can use the great software maps described in the nearby Tip. These tools help you figure out what countries correspond to the call signs you hear. Another handy tool is a prefix list, which helps you figure out call signs' countries of origin. A detailed prefix-country list is available at www.ac6v.com/prefixes.htm#PRI.

While you're collecting resources, here's another suggestion: Centered on your location, an *azimuthal-equidistant* (or *az-eq*) map, such as the one in Figure 11-1, tells you the direction from which a signal is coming. Because signals travel along the "Great Circle" paths between stations (imagine a string stretched tightly around a globe between the stations), the path for any signal you hear follows the radial line from the middle of the map (your location) directly to the other station. If the path goes the long way around, it goes off the edge of the map, which is halfway around the world from your station, and reappears on the other side. Most signal paths stay entirely on the map because they take the short path. Some az-eq maps are available in the *ARRL Operating Manual;* you can also generate a custom map at sites such as `www.wm7d.net/azproj.shtml`.

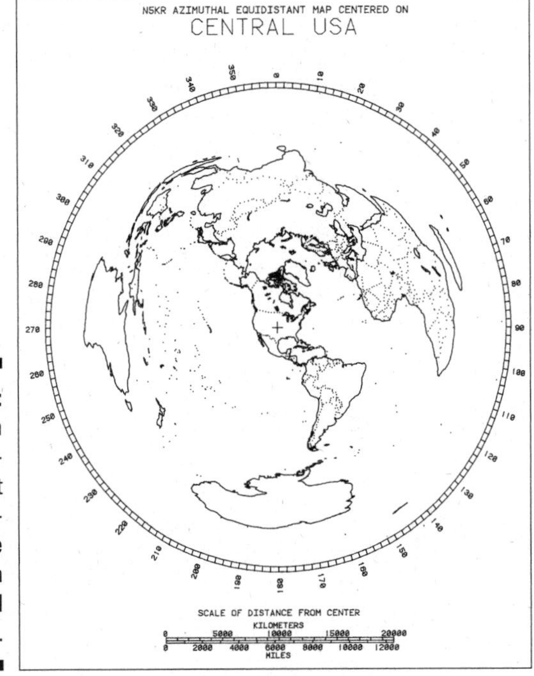

Figure 11-1: An azimuthal-equidistant map centered on the Midwestern United States.

Two great collections of online ham radio maps and mapping software are DX Atlas by Alex Shovkoplyas (VE3NEA), at `www.dxatlas.com`, and Mapability, by Tim Makins (EI8HC), at `www.mapability.com/ei8ic`. Either of these packages is a tremendous asset to ham radio operating and enjoyment.

Daytime signals

You must account for the fluctuations in the ionosphere when you're DXing. Depending on the hour, the ionosphere absorbs a signal or reflects it over the horizon as described in Chapter 8. In the daytime, the 20, 17, 15, 12, and 10 meter bands, called the *high bands,* tend to be open to DX stations — that is, they support propagation. Before daylight, signals begin to appear from the east, beginning with 20 meters and progressing to the higher bands over a few hours. After sunset, the signals linger from the south and west for several hours, with the highest-frequency bands closing first, in reverse order. Daytime DXers tend to follow the *maximum usable frequency* (MUF), the highest signal the ionosphere reflects. These reflections are at a low angle and so can travel the longest distance for a single reflection. (One reflection is called a *hop.*) Because the signal gets to where it's going in the fewest hops, it has a higher signal strength.

Nighttime DXing

From 30 meters down in frequency are the nighttime bands of 30, 40, 60, 80, and 160 meters, known as the *low bands.* These bands are throttled for long-distance communication during the daytime hours by absorption in the lower layers of the ionosphere. At sunset, these bands start to come alive. First, 30, 40, and 60 meters may open in late afternoon and stay open somewhat after sunrise. The 80 and 160 meter bands, however, make fairly rapid transitions around dawn and dusk. Signals between stations operating on 80 and 160 meters often exhibit a short (15- to 30-minute) peak in signal strength when the easternmost stations are close to sunrise, a peak known as *dawn enhancement.* This time is good for stations with modest equipment to be on the air and to take advantage of the stronger signals on these more difficult DX bands.

Tracking the Sun

Because the Sun is so important in determining what bands are open and in what direction, you need to know what portions of the Earth are in daylight and darkness. You can use a variety of tools to keep track of the Sun. The figure below shows the handy map available at `http://dx.qsl.net/propagation/greyline.html`.

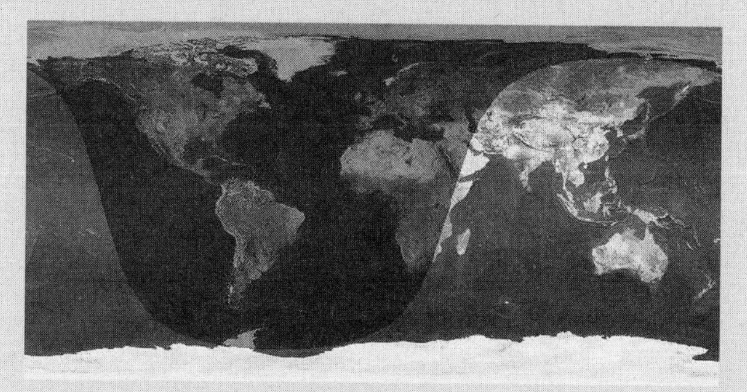

The 160 meter band is known as the *top band* because it has the longest wavelength of any currently authorized amateur band. This long wavelength requires larger antennas. Add more atmospheric noise than at higher frequencies, and you have a challenging situation, which is why some of the most experienced DXers love top-band DXing. Imagine trying to receive a 1-kilowatt broadcast station halfway around the world. That's what top-band DXers are after, and many of them have managed it. More long-wavelength bands are on the way; watch for news about new frequencies to try.

Contacting a DX station

Making a call to a DX station requires a little more attention to the clarity of your speech and the quality of your sending than making a call to a nearby ham does. Your signal likely has the same qualities to the DX station as the DX station's signal does to you — hollow or fluttery and weak — so speak and send extra carefully. Give the other station's call sign, using the same phonetics that the other station is using; then repeat yours at least twice, using standard phonetics. For Morse code or digital contacts, send the DX station's call sign once and your call sign two or three times. The speed of your Morse should be no faster than that of the DX station.

Except when signals are quite strong, DX contacts tend to be shorter than contacts with nearby stations. When signals are weak or the other station is rare, a contact may consist of nothing more than a confirmation that both stations have the call signs correct, as well as a signal report (see Chapter 8). To confirm the contact, both you and the DX station must get each other's call signs correct. To do that, use standard phonetics (on voice transmissions), speak clearly, and enunciate each word.

When it's time to conclude the contact, you need to let the other station know whether you'll confirm the contact by using an online system like Logbook of the World (`www.arrl.org/logbook-of-the-world`) or by sending a QSL card. Collecting QSL cards, like those shown in Figure 11-2, is a wonderful part of the hobby. See Chapter 14 for more info on these cards and on QSLing systems.

YOU DON'T NEED TO SHOUT INTO THE MICROPHONE! Shouting doesn't make you any louder at the other end. By adjusting your microphone gain and speech processor, you can create a very understandable signal at normal voice levels. Save the shouting for celebrating your latest DX contact; your contacts and family members will thank you for doing so.

Figure 11-2: DX QSL cards from New Zealand, Japan, Pratas Island, and England.

If you call and call and can't get through, or if the stations you contact ask for a lot of repeats and fills (in other words, if they often ask you to repeat yourself), you probably have poor audio quality. Have a nearby friend, such as a club member, meet you on the air when the bands are quiet, and do some audio testing. Check to see whether you have hum or noise on your audio. Noise is often the result of a broken microphone cable connection, either in the microphone itself or at the radio connector. The radio's power-meter output alone may not tell you when you have a problem, so an on-the-air check is necessary to find it. Inexpensive, old, and/or noncommunications microphones (such as computer microphones) often have poor fidelity. If your on-the-air friend says that you sound like a bus-station announcer, upgrade to a better microphone.

Navigating pileups

A *pileup* is just that: a pile of many signals trying to get through to a single, often quite rare station. Pileups can sound like real messes, but if you listen carefully, you'll notice that some stations get right through. How do they do this? They listen to the rare station's operating procedure, find what kind of signals the operator is listening for, and carefully time their calls. If they don't get through the first time, they stop calling and listen until they have the pattern figured out. These smooth operators use their ears instead of their lungs or amplifiers to get through. You can, too, by listening first and transmitting second. Here are some common tricks to listen for and try yourself:

- Time your call a little bit differently from everybody else. Wait a second or two before beginning, or wait for the short lull when most hams have given their call signs once and are listening.
- Make your signal sound a little bit different — higher or lower — by offsetting the transmit frequency by 200 Hz or 300 Hz.
- Give your call once or twice before listening for the DX station. Some folks never seem to stop calling; how would they know if the DX station did answer them?
- Use phonetics similar to those of others who have gotten through.
- Try to figure out what the DX station hears well, and do that.

Working split: Split-frequency operation

A *split* refers to a station that's transmitting on one frequency but listening on another. This procedure, called *working split,* is common when many stations are trying to get through to a single station, such as a rare DX station. You can tell that a station is working split when you hear the station contacting other stations but can't hear those stations' responses. The process can also work the other way around: Sometimes, you tune in a pileup of stations trying to contact a DX station, but you aren't able to hear the DX station's responses. Typically, the DX station's split listening frequency is a few kHz above the transmitting frequency. The station being called gives instructions such as "Listening up 2" or "QRZed 14205 to 14210." The former means that the DX station is listening for stations 2 kHz above the transmit frequency; the latter means that the station is listening in the range between the two frequencies (in this case, probably 14.205–14.210 kHz). Your radio's instruction manual can show you how to configure your radio to receive and transmit on different frequencies.

Don't bother trying to spin the dial back and forth between the receive and transmit frequencies; you won't be quick enough. If you aren't experienced with working split, practice with a nearby friend, using low power, until you're comfortable with using your radio's controls that way.

Using (and abusing) spotting networks

DXers share the frequencies and call signs of DX stations that they discover on the air through an extensive worldwide system of websites. A message that describes where you can find the DX station is called a *spot,* and websites that link up to provide and relay the spots form what is called a *spotting network.* Numerous DX spotting websites are listed at `www.ac6v.com/dxlinks.htm`.

Following is an example spot received from the popular DX Summit website (www.dxsummit.fi):

W5VX 7003.7 A61AJ 0142 05 Nov

This spot means that W5VX is hearing A61AJ from the United Arab Emirates (A6 is the prefix of call signs for amateurs in the U.A.E.) on a frequency of 7003.7 kHz at 0142Z (01:42 a.m. in London) on November 5.

Although jumping from spot to spot can be a lot of fun, maintaining your tuning and listening skills is still important. Be sure that you have the station's call sign correct before you put it in your log. It's disappointing to contact what you think is a rare station, only to find out that due to a busted spot, your fabulous DX contact isn't so fabulous after all. Because spotted stations attract quite a crowd, working DX by not chasing the spotted stations and tuning for them yourself may be easier. Don't become dependent on the spotting networks.

Earning awards

Many DXing award programs are available, and the most popular are listed in Table 11-1.

Table 11-1 Popular DX Awards Programs

Sponsor	*Awards Program*	*Achievement*
ARRL (www.arrl.org/awards)	IARU Worked All Continents (WAC)	Make a contact in each of six continents.
	Worked All States (WAS)	Make a contact in each of the 50 U.S. states.
	DX Century Club (DXCC)	Make a contact with 100 of the DXCC entities (currently, 340 countries, islands, and territories).
CQ Magazine (www.cq-amateur-radio.com)	Worked All Zones (WAZ)	Contact all 40 of the world's CQ-defined zones.
	Worked Prefixes (WPX)	Contact stations with different prefixes in the call signs to receive awards.
Radio Society of Great Britain (www.rsgbiota.org)	Islands on the Air (IOTA)	Contact saltwater islands around the world to achieve various levels of awards.

Seeking skimmer spots

Skimmer spots are generated by the CW Skimmer program, available at www.dxatlas.com/cwskimmer. This software uses digital signal processing to receive many Morse code signals at the same time and generate spots automatically. Sometimes, these spots are sent to the spotting networks and show up on your screen. Information from skimmers around the world is collected and made available online through the Reverse Beacon Network (www.reversebeacon.net). As of early 2013, digital-mode skimmers are in the works, and voice-mode skimmers won't be far behind.

Most DX awards programs reward achievement in the same manner. First, you must qualify for the basic award (100 entities, 100 islands, 300 prefixes, and so on). You receive a certificate and your first *endorsement* (a sticker or other adornment that signifies a specific level of achievement). From that point, you can receive additional endorsements for higher levels of achievement: more contacts on one band, more contacts in one geographic region, and so on. For more information on awards, see "Chasing Awards," later in this chapter.

DXing on the VHF and UHF bands

Although DXing on the traditional shortwave bands is popular, an active and growing community enjoys DXing on the bands above 30 MHz. The excitement of extending your station's capability to these bands is being shared by more hams than ever before. The explosion in popularity of VHF/UHF DXing is similar to the explosion of HF DX enthusiasm in the 1960s, when top-quality equipment became available to the average ham. These days, the latest generation of all-band HF/VHF/UHF radio equipment puts top-notch DXing on the shack desktop.

With the exception of the 6 meter band (known as the *magic band* because of its sudden and dramatic openings for distant stations), these higher frequencies usually don't support the kind of long-distance, transoceanic contact that's common on HF, because the ionosphere can't reflect those signals. VHF/UHF DXers look for contacts by using different methods of propagation.

The VHF and UHF bands have undeserved reputations for being limited to line-of-sight contacts because of the limitations of previous generations of relatively insensitive equipment and the prevalence of FM, which takes considerably more signal strength to provide signal quality equivalent to single sideband (SSB) and Morse code transmissions. By taking advantage of well-known modes of radio propagation, you can extend your VHF and UHF range dramatically beyond the horizon.

Finding and working VHF and UHF DX

As on the HF bands, you find DX stations at the lowest frequencies on the band in the so-called weak signal segments. On the 6 meter band, for example, 50.0–50.3 MHz — a 300 kHz segment as large as most HF bands — is where the Morse code and SSB calling frequencies are located. Similar segments exist on all VHF and UHF bands through the lower microwave frequencies.

When you're DXing on VHF or UHF, stay close to these calling frequencies, or set your radio to scan across the low end of the band and leave the radio on. Propagation between widely separated points is often short-lived. If you wait for somebody to call you or e-mail you with news about a DX station, you're probably going to miss the boat. Set your squelch control (squelch mutes the receiver unless a signal exceeds a preset level) so that the radio is barely quieted. If anything shows up on the frequency, the radio springs to life. This way, you (and whoever else is in earshot) don't have to listen to continuous receiver hiss and random noise.

For this type of DXing, I recommend using a small beam antenna (see Chapter 12). A beam antenna is easy to build, is relatively small compared with HF antennas, and is a terrific homebrew project. Mount the antenna for horizontal polarization, with the antenna elements parallel to the ground. You should be able to point the antenna in any horizontal direction, because signals may appear from nearly any direction at any time.

To find out more about VHF/UHF propagation, join one of the many VHF/UHF contest clubs. These helpful, energetic groups make a lot of expertise available through their websites and at meetings. Another strong community of VHF/UHF DXers is in nearly constant communication worldwide at `www.dxworld.com`.

Sporadic E

The term *sporadic E* refers to an interesting property of one of the lower ionospheric layers: the E layer. Somewhere around 65 to 70 miles above the Earth, illumination of the E layer by the Sun produces small, highly ionized regions that are reflective to radio waves — so reflective that they can reflect signals from the 6 meter, 2 meter, 1.25 meter, and (rarely) 70 cm bands back to Earth. These regions, which drift around over the Earth's surface, usually don't last more than an hour or two. While they're available, though, hams can use them as big radio reflectors. Their unpredictable nature has led to the name *sporadic E.*

Sporadic E (or *Es)* propagation occurs throughout the year but is most common in the early summer months and the winter. When sporadic E is present, signals appear to rise out of the noise over a few seconds as the ionized patch moves into position between stations. The path may last for seconds or for hours, with signals typically being very strong in both directions. Working Es with only a few watts and simple antennas is possible. Most VHF and UHF DXers get their start working Es openings on 6 meters; certainly, more people are actively DXing in that way than in any other.

Aurora

Another large ionized structure in the ionosphere is the aurora, which is oriented vertically instead of horizontally like sporadic E but still reflects signals. When a strong aurora is present, it reflects VHF and UHF signals over a wide area.

One of the neatest things about auroral propagation is that it adds its own audible signature to the signals it reflects. If you've ever seen the aurora, you understand how dynamic it is, twisting and shimmering from moment to moment. This movement is even more pronounced for radio waves. The result is that signals reflected by an aurora have a characteristic rasp or buzz impressed on the Morse tone or the spoken voice. A very strong aurora can turn Morse transmissions into bursts of white noise and render voices unintelligible. After you hear the auroral signature, you'll never forget it.

Tropospheric

Tropospheric propagation (also known as *tropo)* occurs in the atmospheric layers closest to the Earth's surface, in an area known as the troposphere. Any kind of large-scale abrupt change in the troposphere, such as temperature inversions or weather fronts, can serve as a conduit for VHF, UHF, and even microwave signals over long distances. If your region has regular cold or warm fronts, you can take advantage of them to reflect or convey your signals.

Tropospheric propagation supports surprisingly regular communications on 2 meters and 1.25 meters, as well as between stations in California and stations on the slopes of Hawaiian volcanoes. A stable temperature-inversion layer forms over the eastern Pacific Ocean most afternoons, so a properly located station on the slope of a volcano at the right altitude can launch signals along the inversion. As the inversion breaks up near land, the signals disperse and are received by mainland amateurs. When conditions are right, mainlanders can send signals back along the same path — more than 2,500 miles!

Meteor scatter

The most fleeting reflectors of all result from the tens of thousands of meteors that enter the Earth's atmosphere each day, traveling at thousands of miles per hour. The friction that occurs as the meteors burn up ionizes the gas molecules for several seconds. These ionized molecules reflect radio signals, so two lucky stations that have the meteor trail between them can communicate for a short period — a minute or so at maximum. The ionized trails reflect radio waves for shorter and shorter durations as frequency increases. As a result, the lowest-frequency VHF band, 6 meters (50 MHz), is the easiest band for beginners to use for making contacts via meteor scatter.

Hams who attempt to make contact in this way are called *ping jockeys,* because the many short reflections off small meteors make a characteristic pinging sound. As you may imagine, ping jockeys go into high gear around the times of meteor showers, large and small. Because of meteor scatter, hams can enjoy meteor showers even during daylight hours.

If you'd like to know more about this unusual mode, check out `www.meteorscatter.org`.

The most common way to make meteor-scatter contacts is to use the suite of software written by one of ham radio's Nobel Prize winners, Joe Taylor (K1JT). Taylor constructed the free WSJT software program (`http://physics.princeton.edu/pulsar/K1JT`) to use sophisticated coding techniques, but ordinary hams can use it with no more than a radio and a computer sound card. WSJT is useful for moonbounce (bouncing signals from one station to another off the Moon), meteor scatter, and other difficult communication tasks.

Mountaintopping

What do you do when all the popular DXing methods fail to provide you an over-the-horizon path? Move your horizon! Because VHF/UHF radios are lightweight and the antennas are small, you can drive, pack, or carry your gear to the tops of buildings, hills, ridges, fire lookouts, and even mountaintops.

The higher the elevation of your station, the farther your signal travels without any assistance from the ionosphere, weather, or interplanetary travelers. Camping, hiking, and driving expeditions can take on a ham radio aspect, even if you're just taking a handheld radio. From the tops of many hills, you can see for miles, and a radio can see even better than you can. These expeditions are particularly popular in VHF contests, discussed in "Taking Part in Radio Contests," later in this chapter. All you have to do is pick up a book of topographic maps of your state, load the car with your radio gear, and head out.

A special award is available for working mountaintop stations. The Summits On the Air (SOTA) program encourages activity by these sociable climbers, who often use ultra-low-power gear, which makes contacts with them challenging and fun. You can read up on SOTA at `www.sota.org.uk`.

Earning VHF and UHF DX awards

Because VHF DX contacts generally aren't as distant or dispersed as their shortwave cousins are, VHF DX awards deal with geographic divisions on a smaller scale, called *grid squares.* Grid squares are the basis for the Maidenhead

Locator System, in which one grid square measures 1° latitude by 2° longitude. Each grid square is labeled with two letters (called the *field)* and two numbers (called the *square)*. A location near St. Louis, Missouri, for example, is in the EM48 grid square. Grid squares are divided even further into subsquares, which are denoted by two additional lowercase letters, as in EM48ss.

Find your grid square by using one of the grid-square lookup utilities listed on the ARRL's Grid Squares page (`www.arrl.org/grid-squares`).

In North America, where countries tend to be large (except in the Caribbean), the primary VHF/UHF award program is the ARRL's VHF/UHF Century Club (VUCC; `www.arrl.org/vucc`). The number of grid squares you need to contact to qualify varies by the band, due to the degree of difficulty. As an example, on the lowest two bands (6 meters and 2 meters) and for contacts made via satellites, contacts with stations in 100 different grid squares are required. The ARRL's Worked All States (WAS) program (refer to Table 11-1, earlier in this chapter) has a vigorous VHF/UHF audience as well.

In Europe, where more countries are within range of conventional VHF/UHF propagation, many of the shortwave DX awards have VHF/UHF counterparts. Many of those awards are based on contacting different countries, too.

Finally, what would DXing be without a distance record? On shortwave bands, with signals bouncing all the way around the world, the maximum terrestrial distance records were set long ago. In VHF/UHF, though, many frontiers are still left. Al Ward (W5LUA) has put together a VHF/UHF/Microwave record list at `www.arrl.org/distance-records`, and new records are added all the time. Maybe your call sign will be there one day.

Taking Part in Radio Contests

If you've never encountered a radio contest before, the concept can seem pretty puzzling. In this section, I clear up any confusion.

Radio contests, or *radiosport,* are competitions between stations to make as many contacts as possible with as many stations as possible within the time period of the contest. Time periods range from a couple of hours to a weekend. Restrictions specify who can contact whom and on what bands, and what information must be exchanged. Often, themes dictate which stations to contact, such as stations in different countries, grids, or states.

When the contest is over, participants submit their logs to the contest sponsor via e-mail or a web page. The sponsor performs the necessary amount of cross-checking between logs to confirm that the claimed contacts actually took place. Then the final scores are computed, and the results are published online and in magazines. Winners receive certificates, plaques, and other nonmonetary prizes.

What's the point of such contests? Well, for one thing, they can be a lot of fun, as many stations are all on the air at the same time, trying for rapid-fire short contacts. In the big international contests, such as the CQ World Wide DX Contest (www.cqww.com), thousands of stations around the world are on the air on the bands from 160 meters through 10 meters. In a few hours, you can find yourself logging a Worked All Continents (WAC) award and being well on your way to earning some of the DX awards I mention earlier in this chapter.

Contests are also great ways to exercise your station and your operating ability to their limits. Test yourself to see whether you can crack contest pileups and copy weak signals through the noise; find out whether your receiver is up to the task of handling strong signals. If you want to increase your Morse code (CW) speed, spend some time in a contest on the CW sub-bands. Just as in physical fitness, competitive activities make staying in shape a lot more fun.

Choosing a contest

Contest styles run the gamut from low-key, take-your-time events occurring on a few frequencies to band-filling events involving hectic activity in all directions. Table 11-2 lists some popular contests.

Table 11-2 Popular Contests

Contest Name	*Sponsor*
ARRL VHF Contests (Jan, June, Sep)	ARRL (www.arrl.org/contests)
ARRL International DX Contest	
ARRL RTTY Roundup	
ARRL Field Day	
ARRL November Sweepstakes	
ARRL 10 and 160 Meter Contests	
CQ World Wide DX and WPX Contests	*CQ* Magazine (www.cq-amateur-radio.com)
IARU HF Championship	IARU (www.arrl.org/contests)
North American QSO Parties	National Contest Journal (www.ncjweb.com)
Worked All Europe (WAE)	DARC (www.darc.de/referate/dx/contest/waedc/en)

Most contests run annually, occurring on the same weekend every year. The full-weekend contests generally start at 0000Z (Friday night in the United States) and end 48 hours later (Sunday night in the United States) at 2359Z. You don't have to stay up for two days, but some amazing operators do. Most contests have time limits or much shorter hours.

Start by finding out what contests are coming up. Use Table 11-3 to locate several sources of information, or enter *contest calendar* in a web search engine. The list of contests will also help you identify which contest a station is participating in if you hear it on the air. Most websites include the contest rules or a link to the contest sponsor's website.

Table 11-3 Contest Calendars

Calendar	*URL*
ARRL Contest Calendar	www.arrl.org/contests/calendar.html
ARRL Contest Update (biweekly e-mail newsletter, free to ARRL members)	www.arrl.org/contest-update-issues
SM3CER Contest Service	www.sk3bg.se/contest
WA7BNM Contest Calendar	www.hornucopia.com/contestcal

After you know the rules for a particular contest, listen to a participating station. The most important part of each contact is the information passed between stations, known as the *exchange*. For most contests, the exchange is short — a signal report and some identification such as a *serial number* (the count of contacts you made), name, location, or club membership number. By reading the rules or simply listening, you'll know what's required and the order in which you need to send your information. When you're ready, give the contest a try.

If you don't know the rules of a contest but want to help a station calling "CQ contest" with a contact, wait until the station doesn't have anyone calling and then ask what information is needed. Stations in the contest want your contact and generally guide you through whatever they need.

Operating in a contest

Don't be intimidated by the rapid-fire action that occurs during contests. Contesting is unusual as a sport, in that the participants score by cooperating with one another. Even archrivals need to put each other in their logs to earn points. All the participants, including the big guns, need and want to talk to you.

You needn't have a huge and powerful station to enjoy contesting; most contesters have a simple setup. Besides, the most important part is the operator. If you listen, know the rules, and have your station ready to go, you're all set.

Making contest contacts

Here's an example of a contact in a typical contest: the Washington State Salmon Run. (State contests are often referred to as *QSO parties,* to emphasize their easygoing style.) In this scenario, I'm W7VMI in King County, calling CQ to solicit contacts, and you're W1AW in Connecticut, tuning around the band to find Washington stations. The information we exchange is a signal report (see Chapter 8) and my county and your state, because (at least for this example) you're not in Washington.

> **Me (W7VMI):** CQ Salmon Run CQ Salmon Run from Whiskey Seven Victor Mike India.
>
> **You (W1AW):** Whiskey One Alfa Whiskey.
>
> (***Note:*** You send or say your call sign once, using standard phonetics on voice transmissions.)
>
> **W7VMI:** W1AW, you're five-nine in King County.
>
> **W1AW:** QSL, W7VMI, you're five-nine in Connecticut.
>
> **W7VMI:** Thanks, QRZed Salmon Run Whiskey Seven Victor Mike India.

The whole thing takes about ten seconds. Each station identifies and exchanges the required information. "Five-nine" is the required signal report signifying "loud and clear." That's an efficient contest contact, and most contacts are similar.

When the contact is over, keep tuning for another station calling "CQ contest." This method of finding stations to call is *searching and pouncing,* which I cover later in this chapter.

What if you miss something? Maybe you've just tuned in the station, and the band is noisy or the signal is weak. To continue the preceding example, my response to your call might sound like this:

> **Me (W7VMI):** W1AW, you're five-nine in BZZZZTCRASH@#$%^&*.
>
> **You (W1AW):** Sorry, please repeat your county.
>
> **W7VMI:** Kilo India November Golf, King County.
>
> **W1AW:** QSL, W7VMI, you're five-nine in Connecticut.

You're probably thinking, "But I missed the county. How can the signal report be five-nine?" By convention, most contesters say "Five-nine," type 599, or send 5NN in Morse code (the *N* represents an abbreviated 9). The signal report doesn't affect the score unless it's miscopied.

Contesting is no more complicated than getting your sandwich order taken at a busy deli counter during lunch hour. Contesting has a million variations, but you'll quickly recognize the basic format.

If you're unsure of yourself, try "singing along" without actually transmitting. Make a cue card that contains all the information you need to say or send. If you think you may get flustered when the other station answers your call, listen to a few contacts, and copy the information ahead of time. Serial numbers advance one at a time, so you can have all the information before your contact.

Your score for almost all contests is made up of QSO points and multipliers. Each contact counts for one or more QSO points, sometimes depending on the mode, band, or other special consideration. Multipliers — so named because they multiply QSO points for the final score — are what make each contest an exciting treasure hunt. Depending on the theme, you may be hunting for states, grids, counties, lighthouses, islands, or anything else. Read the rules carefully to find out how the multipliers are counted: only once, once per band, once per mode, and so on. Special bonus points may be awarded for working certain stations or multipliers.

You don't have to be a speed demon; just be steady. Good contest operators are smooth and efficient, so send your full call sign once. If the station answers with your call sign, log the exchange and send your information only once, even if you're using a small station. The other operator will ask you to repeat yourself if he or she misses some of the information.

Logging your contacts

Manual logging (with pencil and paper) is the easiest method when you're a beginner. Often, the contest sponsor has a log sheet that you can download or print from a website. After the contest, you can convert your written entry to electronic form by using logging software or an online converter such the one at `www.b4h.net/cabforms`.

If you're a more experienced contester, using a general-purpose logging program or special contesting software makes contesting much easier. The software keeps score, maintains a *dupe list* (a list of stations you've already worked), shows needed multipliers, connects to spotting networks, and creates properly formatted logs to submit to the sponsors.

Table 11-4 lists some popular software programs. Entering *contest logger* in a search engine also turns up many useful programs.

Table 11-4 Popular Contest Logging Software

Software	*URL*
CQ/X (for mobile operation in state QSO parties)	`www.no5w.com`
N1MM Logger	`http://n1mm.hamdocs.com/tiki-index.php`
N3FJP contest loggers	`www.n3fjp.com`
Rover Log	`http://code.google.com/p/roverlog`
Win-Test	`www.win-test.com`
WriteLog	`www.writelog.com`

Most contests expect logs in *Cabrillo format,* which is nothing more than a method of arranging the information in your log so that the sponsor's log-checking software can read it. To find out more about Cabrillo, visit `www.arrl.org/cabrillo-format-tutorial`.

Many sponsors post a Logs Received web page that you can check to make sure that your log was received. Don't miss the deadline — usually a few days after the contest — for submitting logs. Even if you're not interested in having your score posted in the results, submitting your log just for the sponsor to use in checking other logs helps improve the quality of the final scoring.

Taking tips from winners

After you've participated a few contests, you may feel that the top scores are out of your reach. How do the contest winners achieve them? They'll tell you that no magic is involved: Winning contests comes down to perseverance and patient practice. The following sections discuss a few tricks of the trade that you'll develop with time.

Calling CQ in a contest

To make a lot of contacts, you have to call CQ. In any contest, more stations are tuning than calling. You can turn those numbers to your advantage. Find a clear frequency (see "Being polite," later in this chapter), and when you're sure that it's not in use, fire away.

Following are a few examples of appropriate ways to call CQ in a contest. (Replace *Contest* with the name of the contest or an abbreviation of the name.)

- **Voice transmissions:** CQ Contest CQ Contest from Whiskey One Alpha Whiskey, Whiskey One Alfa Whiskey, Contest.
- **Morse code or digital modes:** CQ CQ TEST DE W1AW W1AW TEST.
- **VHF/UHF transmissions:** CQ Contest from W1AW grid FN31.

Keep transmissions short, and call at a speed at which you feel comfortable receiving a reply. Pause for two or three seconds between CQs before calling again. Other stations are tuning the band and can miss your call if you leave too much time between CQs.

When you get a stream of callers going, keep things moving steadily. Try to send the exchange the same way every time. On voice, don't say "uh" or "um." Take a breath before the exchange, and say everything in one smooth sentence. As you make more contacts, your confidence builds. An efficient rhythm increases your *rate* — the number of contacts per minute.

Contesting being what it is, you'll eventually encounter interference or a station that begins calling CQ on your frequency. You have two options: stick it out or move. Sometimes, sending a simple "The frequency is in use, CQ contest . . ." or "PSE QSY" (which means "Please change your frequency") on CW and digital modes does the trick. Otherwise, unless you're confident that you have a strong signal and good technique, finding a new frequency may be more effective. The high end of the bands is often less crowded, and you may be able to hold a frequency longer.

Searching and pouncing

Searching and pouncing (S&P) is usually accomplished by tuning across the band and finding stations manually. Another popular method is to connect to the spotting network and use logging software to create a list of stations and their frequencies, called a *band map*. If your logging software can control your radio, all you have to do is click the call signs to jump right to their frequencies.

If you call and get through right away, terrific. Sometimes, though, you won't get through right away. Use your radio's memories or alternate variable-frequency oscillator (VFO; see Chapter 8). By saving the frequencies of two or three stations, you can bounce back and forth among several pileups and dramatically improve your rate.

Many stations use the spotting networks to find rare or needed stations in a contest. Be aware that using such information usually requires you to enter in an assisted or multiple-operator category. Know the rules of the contest regarding spotting information, and be sure to submit your score and log in the proper category.

If you do use information from the spotting networks, don't assume that the call sign is correct. Always listen to make sure that it's correct, because many spots are incorrect *(busted)*. If you log the wrong call sign, you'll not only lose credit for the contact, but also incur a small scoring penalty.

Being polite

Large contests can fill up most or all of an HF band, particularly during voice-mode contests, and often cause friction with noncontest operators. As in most conflicts, each side needs to engage in some give-and-take to keep the peace. If you're participating in a contest, be courteous, and make reasonable accommodations for noncontesters. If you're not contesting, recognize that large competitive events are legitimate activities and that you need to be flexible in your operating expectations.

That said, how can you get along with everyone? Here are a few tips:

- **Make sure that your signal is clean.** *Clean,* in this context, means not generating key clicks or splatter from overmodulation. (You may hear about such problems from stations operating near you.) A distorted signal's intelligibility is greatly reduced. A clean signal gets more callers every time and occupies less bandwidth.
- **Make sure that your receiver isn't overloaded.** Keep your noise blanker and preamp off (read about these devices in your radio's operating manual), and use every receiver adjustment on the front panel, including the front-end attenuator.
- **Listen before you leap.** Noncontest contacts are relaxed, with long pauses, so a couple of seconds of dead air don't mean that the frequency is clear. Asking "Is the frequency in use?" (QRL? in Morse code) before calling CQ is the right thing to do, whether you're in a contest or not. If a Morse code contact is ongoing, the response to your query may be a dit (meaning "Yes, it's busy") if the other operator is in the middle of trying to copy an exchange from a different station.

When you're participating in a contest, keep a minimum of 1.5 kHz between you and adjacent contest contacts on phone and 400 Hz on CW or radioteletype (RTTY). Don't expect a perfectly clear channel. Contesters should tune higher in the band to find less-congested frequencies and give noncontest QSOs a wider margin.

- **Avoid major net frequencies.** Examples include the Maritime Service Net on 14.300 MHz. Also, be aware of any emergency communications declarations or locations where regional emergency nets meet, and give those frequencies a wide berth. Those frequencies are often busy with noncontest activity.

Chasing Awards

If the awards I mention in the section on DXing pique your interest, the following sections may satisfy that interest by discussing awards in greater detail. Seeking awards is one of the most fulfilling activities in the hobby of ham radio. Certificates (often called *wallpaper*) are the usual rewards. Some radio shacks that I've visited are literally papered (ceilings, too) with certificates and awards in all shapes, sizes, and colors. Some wallpaper is plain, but it can be as colorful and as detailed as paintings or photographs.

If awards sound interesting, you may be a member of the species of ham known as the paper chaser or wallpaper hanger. Believe me, a lot of them are out there!

Finding awards and special events

You can find awards almost everywhere you look. *CQ* magazine, for example, runs a column featuring novel awards every month, and The K1BV DX Awards Directory (`www.dxawards.com`) lists more than 3,300 awards from nearly every country. Want to try for the "Tasmanian Devil" award? Contact VK7 (Tasmania's prefix) amateurs. Or pursue the South African Relay League's "All Africa" award for contacting the six South African call areas and 25 other African countries.

Most awards have no time limit, but some span a given period, often a year. Whatever your tastes and capabilities, you can find awards that suit you (see Figure 11-3). You can find the awards directory and an extensive list of web links, along with the K1BV directory, at `www.dxawards.com`.

Along with ongoing awards programs, you can find many special-event stations and operations, which often feature special call signs with unusual prefixes (of great interest to hams who chase the WPX award; refer to Table 11-1, earlier in this chapter) and colorful, unusual QSL cards. Ham stations often take part in large sporting events and public festivals, such as international expositions and the Olympic Games. The larger special-event stations are well publicized and are listed on web pages such as `www.arrl.org/special-event-stations` and on the ham Internet portals I discuss in Chapter 3. Other special-event stations just show up unannounced on the air, which makes finding them exciting.

Finding out more about contesting

The best way to find out more about contesting is to work with an experienced contester. You'll probably find that one or two multiple-operator stations in your region are active in the big contests. Look through the results of previous contests for their call signs. Contact the station owner, and volunteer to help out; most operators are eager to have you on board or can help you find another team. As a rookie, expect to listen, log, or spot new multipliers, all of which are valuable learning opportunities. When you know the ropes, you can fill in on the air more and more.

You can also find many contest clubs around the country. Look at the club scores in the results, and contact them.

QST and *CQ* magazines feature contest results and articles on technique. The ARRL also publishes the *National Contest Journal* (`www.ncjweb.com`), which sponsors several HF contests every year and features interviews with contesters, as well as articles and columns on contesting. ARRL members can receive the biweekly e-mail newsletter Contest Update (`www.arrl.org/contest-update-issues`) without charge. The CQ-Contest e-mail list is a source of many good ideas; subscribe to it at `http://lists.contesting.com/mailman/listinfo/CQ-Contest`.

Figure 11-3: Participating in a contest can result in an attractive certificate.

Recording qualifying contacts

Before embarking on a big adventure to achieve an obscure award, find out whether the award is still active by checking with the sponsors. Get a positive "go ahead" if you have the slightest question whether an award program is active.

Determine whether the award requires you to submit QSL cards. Overseas sponsors may allow you to submit a simple list of contacts that comply with their General Certification Rules (GCR) instead of requiring you to submit actual cards.

When you make an eligible contact, be sure to log any information that the award may require. If you're working Japanese cities, for example, some awards may require you to get a city number or other ID. This information may be printed on the QSL card that the other station sends you, so ask for the information during the contact itself. Grid-square information isn't always on QSLs, either. Be sure to ask during the contact or in a written note on the QSL you send to the other station.

If you make a contact for an award that favors a certain geographic area, ask the station's operator whether he or she will let others know you're chasing the award, which may even generate a couple more contacts for you right on the spot. Certainly, the request lets stations in the desired area know to listen for you on the band or perhaps arrange a schedule. This technique helps a lot for difficult awards or contacts in remote areas.

Applying for awards

When applying for an award, use the proper forms, addresses, and forms of payments. Follow the instructions for submitting your application to the letter. If you aren't certain, ask the sponsor. Don't send your hard-earned QSLs or money before you know what to do.

When you do apply for the awards, you may want to send the application by registered or certified mail, particularly if precious QSLs or an application fee are inside. Outside the developed countries, postal workers are notorious for opening any mail that may contain valuables. Make your mail look as boring and ordinary as possible; keep envelopes thin, flat, and opaque.

Mastering Morse Code

Mastering Morse code is a personal thing, such as playing an instrument or achieving a new athletic maneuver. Many people liken it to studying another language because it involves the same sudden breakthroughs after periods of repetition. Becoming a skilled Morse code operator results in a great sense of accomplishment, and you'll never regret taking the time to learn it.

If you decide to study Morse code, some methods are much better than others. Avoid any method that encourages you to think of a table of character patterns. Managing the required five words per minute while looking up each character in your mind is difficult. After you can receive code as fast as these methods permit, you'll find it hard to move to the higher speeds that make Morse code fun.

Starting with Farnsworth

The style that most hams are successful with is the *Farnsworth method.* The dits and dahs of each character are sent at the code speed you want to achieve (measured in words per minute [wpm]), but the individual characters are spaced far enough apart in time that the overall word speed is low enough for you to process the character's sound pattern. This spacing is called *character spacing.* For the beginner, a common sending speed for individual characters is 17 wpm; the character spacing results in a much lower overall speed for the words.

By sending the character elements at high speed, you hear them as one continuous pattern and keep from falling into the look-it-up-in-the-table trap. Thinking of the code as a table of letters in one column and a dot-dash pattern in another is a natural tendency. When you hear a sequence of code elements (the dots and dashes), such as short-long-long, you look it up in your head to find the character *W.* This method works, but only up to speeds of a few wpm, and it's a very hard habit to break.

Choose a study aid that uses the Farnsworth or Koch method. The ARRL and Gordon West study tapes and books are good choices, as are the Ham University and Morse Academy software. (Go to `www.ac6v.com/morseprograms.htm` for an encyclopedic listing of Morse code training aids.) The FISTS Club (`www.fists.org`), a group dedicated to helping hams learn Morse code, also offers low-cost training software and on-the-air assistance and training.

Sharpening your skills

Figure 11-4 shows the normal progression in code speed. Between steps, or plateaus, you achieve new skills. While you're on a plateau, you refine or solidify the skill. Over time, you progress from copying letter by letter to hearing whole groups of characters and then words.

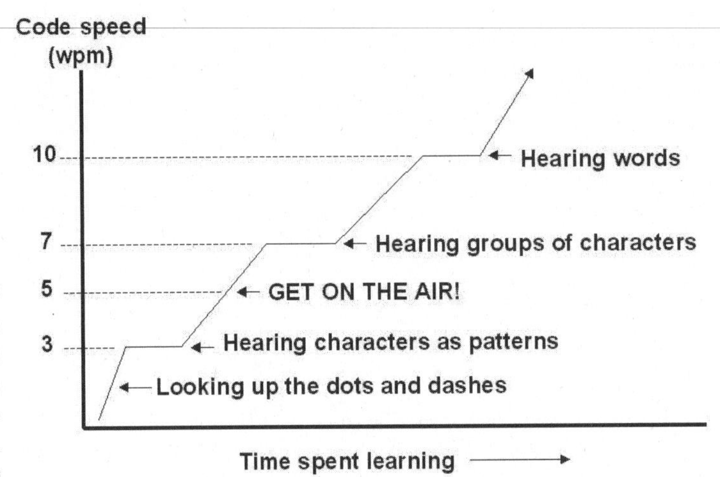

Figure 11-4: The normal progression in code speed.

A great way to gauge your Morse code proficiency is with live code practice, which enhances your taped or computer-generated studies. The most widely received code-practice sessions are transmitted by the ARRL's station W1AW in Newington, Connecticut (`www.arrl.org/w1aw`). Code practice may be available on a VHF or UHF repeater in your area, too. Check with your local radio clubs to find out.

The secret of mastering Morse code is keeping at it. You'll have days when conquering new letters and higher speeds seem to come effortlessly; then you'll have days when progress is elusive. Those plateau days are the most important times to keep going, because they're the times when your brain is completing its new wiring.

As you discover more code, work it into everyday life. While you're driving to work, for example, whistle or hum the code for license plates, billboards, and signs.

Soon, you'll effortlessly copy bits and pieces that seemed impossible to grasp only days before. Characters that were hopelessly opaque become as natural as speech. Trust me — learning Morse code well enough to begin making contacts is within your grasp if you're willing to give it a try.

Operating at Low Power (QRP)

Why would you want to use low power and a weak signal instead of high power and a strong signal? Skill. Putting as little as possible between yourself and the station at the other end and still making the contact takes skill. Build up a little experience and then give QRP a try.

QRP is up to 5 watts of transmitter output power on Morse code or digital transmissions and 10 watts of peak power on voice, usually SSB. The quality of your antenna or location isn't considered, just transmitter power. If you choose to turn the power down below 1 watt, you're *milliwatting*.

QRP is primarily an HF activity, and most QRP contacts are in Morse code due to the efficiency and simplicity of that mode. QRPers often hang out around their calling frequencies, shown in Table 11-5. For digital modes with QRP, use the digital calling frequencies listed later in this chapter.

Table 11-5 North American QRP HF Calling Frequencies

Band (Meters)	*Morse Code (MHz)*	*Voice (MHz)*
160	1.810	1.910
80	3.560, 3.710	3.985
40	7.030, 7.110	7.285
30	10.106	
20	14.060	14.285
17	18.096	
15	21.060, 21.110	21.385
12	24.906	
10	28.060, 28.110	28.385, 28.885

Getting started with QRP

To start QRPing, just tune to a clear frequency nearby and call CQ. You don't need to call CQ QRP unless you specifically want to contact other QRPers.

If you're just getting started, tune in a strong signal, and give that station a call with your transmitter output power turned down to QRP levels. (Check your radio operating manual for instructions.) Make sure that your transmissions are clear, which allows the other station to copy your call sign easily.

Some low-power stations send their call with /QRP tacked onto the end to indicate that they're running low power. This procedure isn't necessary and can be confusing if your signal is weak. After all, that's more characters for the other station to copy, isn't it?

Getting deeper into QRP

When you have some experience with QRPing, you may want to get deeper into that aspect of the hobby by taking up the following pursuits:

- **Building your own QRP gear:** Many QRPers delight in building their own equipment — the smaller and lighter, the better. You can find lots of kits, such as the popular backpackable PFR-3 transceiver from Hendricks QRP Kits (`www.qrpkits.com`), shown in Figure 11-5, and homebrew designs for hams who have good construction skills.

  QRPers probably build more equipment than those in any other segment of the hobby, so if you want to find out about radio electronics, you might consider joining a QRP club (discussed later in this list) and one of the QRP e-mail mailing lists.

- **Entering QRP contests:** You'll find some QRP-only contests, and nearly all the major contests have QRP categories. Many awards have a special endorsement for one-way and two-way QRP. The QRP clubs themselves have their own awards, including my all-time favorite, the 1,000 Miles Per Watt award (`www.qrparci.org`). Some stations make contact with so little power that their contacts equate to millions of miles per watt!
- **Joining QRP organizations:** QRPers are enthusiastic and helpful types, always ready to act as QRP Elmers. Their clubs and magazines are full of "can-do" ham spirit.

Table 11-6 lists some QRP resources, including large QRP organizations and online groups.

Figure 11-5: The three-band Pack-Friendly Radio (PFR-3) QRP CW transceiver.

Table 11-6 QRP Operating Resources

Organization	*URL*	*Resources/ Information*
Adventure Radio Society	`www.arsqrp.blogspot.com`	Portable operation
eHam.net QRP forums	`www.eham.net/ehamforum/smf/index.php` `www.eham.net/ehamforum/smf/index.php/board,23.0.html`	Discussions on a wide variety of topics
GQRP Club	`www.gqrp.com`	*Sprat* magazine, and building and operating information
QRP Amateur Radio Club International (ARCI)	`www.qrparci.org`	*QRP Quarterly* magazine and numerous awards
QRP-L e-mail reflector	`www.mailman.qth.net/mailman/listinfo/qrp-l`	Best-known QRP e-mail reflector; archives for e-mail, files, and articles

Look for special gatherings of QRP enthusiasts at conventions, such as the internationally attended Four Days in May (www.qrparci.org/fdim), which coincides with the Dayton Hamvention (see Chapter 3). Other QRP gatherings occur around the United States throughout the year.

Getting Digital

Operating via digital modes is the fastest-growing segment of amateur radio today. Applying the power of digital signal processing (DSP) allows hams to communicate keyboard to keyboard, using just a few watts of output power and modest equipment.

On HF bands, digital modes must overcome the hostile effect of the ionosphere and atmospheric noise on delicate bits and bytes. Modes such as PSK31, PACTOR, and Throb make short transmissions with robust error detection and correction mechanisms. Limits on transmission bandwidth keep data transmission rates to less than 10 Kbps, but these restrictions have stimulated hams to create interesting protocols.

On VHF bands, digital data modes have fewer restrictions on bandwidth. The bands are quieter, with less fading and interference, so data communication works better. At 440 MHz and up, you can use 56 kilobaud technology, and the D-STAR system supports an Ethernet bridge mode that connects two computers via the amateur 1.2 GHz band as though they were connected by a standard cable.

You'll find a selection of articles about modes at www.arrl.org/digital-data-modes. Also, the free multiple-mode software FLDIGI (see Chapter 8) allows you to tune in and try many modes.

PSK modes

The most widely used digital mode on the HF bands is PSK31. Peter Martinez (G3PLX) invented it — a great example of ham innovation — and developed a complete package of Windows-based software to support it. He generously placed his creation in the ham radio public domain, and hams are adopted it like wildfire. If you want to find out more about using PSK31, Table 11-7 contains several good resources. Also, the PODXS Ø7Ø Club (www.podxs070.com) specializes in this popular mode.

Table 11-7 PSK Resources

Resource	*Description*
DigiPan (www.digipan.net)	Free software for operating PSK31 and PSK63
FLDIGI (www.w1hkj.com/Fldigi.html)	Free software that supports a large number of digital modes
PSK31 (www.aintel.bi.ehu.es/psk31.html)	Latest updates on the mode and software, and an e-mail reflector for discussing PSK31
PSK31 tutorial (www.podxs070.com/introduction-to-psk)	Introduction to PSK31, explaining how it works and how to operate on the mode

PSK stands for *phase shift keying,* and *31* represents the 31.25-baud rate of the signal — about regular typing speed. It also uses a new coding system for characters, called *Varicode,* that has a different number of bits for different characters, not unlike Morse code. Instead of a carrier turning on and off to transmit the code, a continuous tone signifies the bits of the code by shifting its timing relationship (known as *phase)* with a reference signal. A receiver syncs with the transmitter and decodes even very noisy signals because the receiver knows when to look for the phase changes.

PSK31 is very tolerant of the noise and other disturbances on HF bands. In fact, you can obtain nearly solid copy with signals barely stronger than the noise itself. Figure 11-6 shows a DigiPan software display of several signals, some of which are quite weak. (The screen shot of FLDIGI in Chapter 8 shows another example of PSK31 reception.) The lighter streaks represent signals, and each horizontal line represents a new sampling of the receiver's output. New signals appear at the top and slowly drift downward in a waterfall display.

Because the bandwidth of PSK31 is so narrow, finding other PSK31 stations on the air requires a pretty good idea of where they are. The most common frequencies are 3580.150, 7035.0, 10142.150, 14070.150, 18100.150, and 21080.150 MHz.

Since PSK31's introduction, several enhancements have been made in the original protocol, including a variant called PSK63 that adds some features and quality improvements at the expense of bandwidth.

PSK31 contests are held everywhere, and a growing community of PSK31/63 users exists around the world.

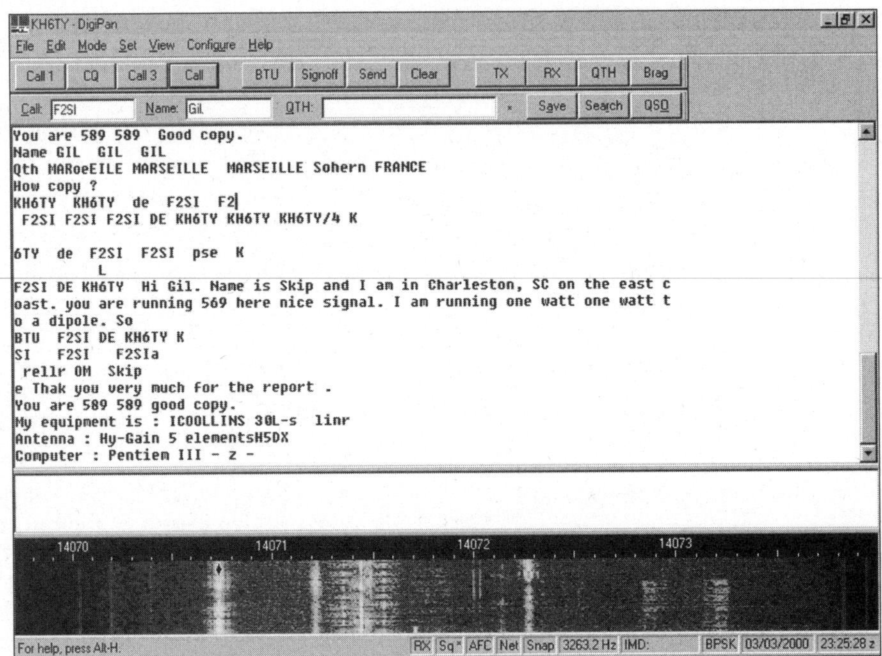

Figure 11-6: A PSK31 QSO using the waterfall display of DigiPan.

Radioteletype

Morse code was the first real digital mode, but the first fully automated data transmission protocol was radioteletype (RTTY). Commercialized in the 1930s, RTTY (pronounced *ritty* by hams) uses a 5-bit code known as Baudot — the origin of the word *baud.* The Baudot code sends plain-text characters as 5-bit codes that use alternating patterns of two audio frequencies known as *mark* and *space,* creating a type of modulation called *frequency shift keying* (FSK).

The tones 2125 Hz (mark) and 2295 Hz (space) fit within a normal voice's bandwidth, so the RTTY signal can be transmitted with a regular SSB transceiver in place of a voice signal. On the receiving end, the SSB transceiver receives the transmission as an audio signal. The text characters can be recovered from the pair of mark and space tones by an external decoder or a computer and sound card running software such as MMTTY (see Table 11-8).

Fans of the antique teleprinters that used rotating mechanical contacts and briefcase-size tone encoding equipment keep them running and use them on the bands even today. If you get a chance to watch one of these devices at work, you'll be amazed by its mechanical complexity.

Although RTTY is being supplanted by the more modern modes, it's still a strong presence on the bands. Tune through the digital signals above the CW stations, and you'll hear lots of two-tone signals "diddling" to each other. A sizable community of RTTY DXers and several major award programs have RTTY endorsements. DXpeditions (refer to "DXing," earlier in this chapter) often include RTTY in their operating plans as well. Table 11-8 lists several online resources for beginning RTTY operators.

Table 11-8 RTTY Resources

Resource	*Description*
AA5AU's RTTY page (`www.aa5au.com/rtty.html`)	Tutorial information, links to RTTY programs, troubleshooting, and RTTY contesting
MMTTY software (`http://hamsoft.ca/pages/mmtty.php`)	Most popular RTTY software for computers
RTTY e-mail reflector (`http://lists.contesting.com/mailman/listinfo/rtty`)	International membership e-mail group

PACTOR and WINMOR

A user of RTTY (refer to the preceding section) quickly discovers that the fading and distortion common on HF can do serious damage to characters sent via Baudot code. Teleprinting over Radio (TOR) enables radio systems to send text characters. TOR systems include data organization and error-correction mechanisms to overcome the limitations of RTTY.

PACTOR goes one step further by adding error-checking packets to the mode. PACTOR II and III add error correction. PACTOR adjusts its speed based on conditions as well. PACTOR III is the most recent release of this technology available to hams. PACTOR III and subsequent versions are available only in equipment available from SCS (`www.scs-ptc.com/pactor`).

Rick Meuthing (KN6KB) developed WINMOR mode as a nonproprietary alternative to PACTOR III communications. WINMOR achieves almost the same data rates as the advanced PACTOR modes and can be used with a computer and sound card; no external controller is required. WINMOR is commonly used in the Winlink e-mail system, discussed in Chapter 9.

Similar data modes include the proprietary family of CLOVER modes developed by HAL Communications. These modes use transmitted waveform shapes and frequencies that are carefully managed to keep the signal within a 500 Hz bandwidth and decrease errors caused by HF propagation.

Packet

Packet is short for *packet radio,* a radio-based networking system based on the commercial X.25 data transfer protocol. Developed by the Tucson Amateur Packet Radio group (TAPR), packet can send error-corrected data over VHF links, which led to the creation of novel data systems for hams. You can read more about packet and one of its best-known applications, APRS, at `www.tapr.org/packetradio.html`.

With packet, ordinary VHF/UHF FM transceivers transfer data as audio tones. An external modem called a terminal node controller (TNC) provides the interface between the radio and a computer or terminal. Data is sent at 1200 or 9600 baud as packets of variable length up to about 1,000 bytes. The protocol that controls packet construction, transmission control, and error correction is called AX.25 (for Amateur X.25). Some packet systems can also use the TCP/IP Internet protocol.

Like wired networks, packet systems are connected in many ways. A packet controller is called a *node.* The connection between nodes is a *link.* Connecting to a remote node by using an intermediate node to relay packets is called *digipeating.* A node that does nothing but relay packets is a *digipeater.* A node that makes a connection between two packet networks or between a packet network and the Internet is called a *gateway.* (The APRS system, described later in this chapter, uses packet networks to relay its information.)

MFSK modes

When you tune around the digital signal areas of the bands, you're likely to come across signals that sound like crazy calliopes or steam whistles playing what sound like random melodies. These signals are *multiple frequency shift keying* (MFSK) modes. Instead of varying the phase or amplitude of a signal to carry the information, MFSK uses different tones or sequences of tones to do the job. Because the data is digital 1s and 0s, the mode is made up of a group of discrete tones that you hear as "notes" on your receiver.

OLIVIA and MFSK16 are the most popular MFSK modes at this writing, but hams are innovating like crazy, and new modes are popping up all the time.

The FLDIGI software package, mentioned earlier in this chapter, stays current with the variations as they come along. You can find the technical details on MFSK modulation at www.qsl.net/zl1bpu/MFSK.

Amateur WLAN and high-speed data

The widespread adoption of wireless local area network (WLAN) technology such as Wi-Fi has brought the same technology within reach of hams as well. Hams share portions of the 2.4 GHz and other microwave bands with unlicensed LAN devices but without the power restrictions of consumer equipment.

Ham experimenters have adapted commercial WLAN protocols, with higher transmitter power and larger antennas than commercial technology products. The largest group is Broadband-Hamnet (http://hsmm-mesh.org). Other groups, such as HamWAN (https://www.hamwan.org), are building regional networks.

Some hams are also using spread-spectrum modems on the UHF and microwave ham bands. At present, a relatively small community of experimenters is attempting to extend the range of commercial technology on amateur frequencies. TAPR hosts the largest community of spread-spectrum experimenters; for information, visit www.tapr.org/spread_spectrum.html.

Digital voice (D-STAR and Codec2)

Amateur radio is also using digital voice links. Icom, one of the largest radio manufacturers, has introduced a D-STAR system of radios, repeaters, and networking equipment (discussed in Chapter 9).

The digital data in the low-speed D-STAR transmissions includes both digitized voice and data characters, which has fostered applications such as D-RATS keyboard-to-keyboard mode. (A high-speed Ethernet bridge mode of D-STAR operates on the 1.2 GHz band.) The D-STAR protocols allow simultaneous voice and data use, leading to variations on text messaging, interfaces to the APRS network (described later in this chapter), and slow-scan TV image transfers, among other uses.

Codec2 is a new digital voice codec designed by David Rowe (VK5DGR) to work well on HF's noisy frequencies as well as on the better-behaved VHF and UHF bands. Codec2 is an open-source software project, so expect to see it adapted in many ways. You can keep track of its progress at www.codec2.org.

APRS

The Automatic Packet Reporting System (APRS), an amateur invention that marries GPS positioning and packet radio, was developed by Bob Bruninga (WB4APR).

The most common use of APRS is to relay location data from GPS receivers via 2 meter radio as an "I am here" service. APRS packets are received directly by other hams or by a packet radio digipeater. If they're received by a digipeater, the packets may also be relayed to a gateway station that forwards the call-sign and position information to an APRS server accessed through the Internet. After the information is received by a server, it can be viewed through a web browser or an APRS viewing program.

The map in Figure 11-7 shows the location of WA1LOU-8. WA1LOU is a call sign, and -8 is a secondary station ID (SSID), allowing the call sign to be used for several purposes with different SSIDs. The figure shows that WA1LOU is in Connecticut, 7.9 miles northeast of Waterbury. You can zoom in on WA1LOU's location. If the radio changes location, this change is updated on the map at the rate at which the operator decides to have his APRS system broadcast the information.

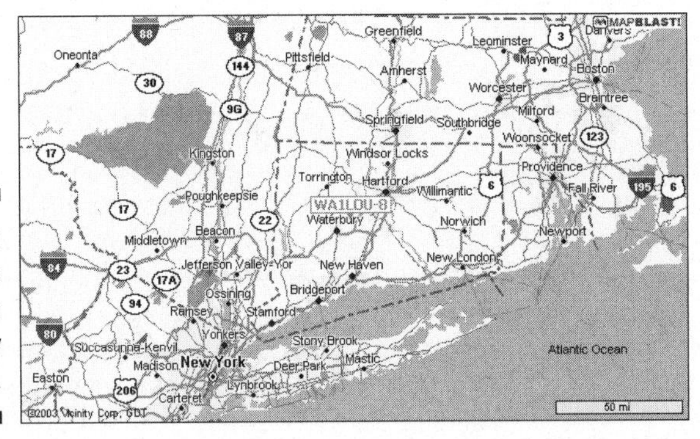

Figure 11-7: WA1LOU's position reported by APRS.

If you have a GPS receiver with an NMEA (National Marine Electronics Association) 0183 data output port and a 2 meter rig and packet radio TNC, you're ready to participate. (Kenwood and Yaesu even make APRS-ready 2 meter radios that include GPS receivers or have direct GPS data interfaces.) The most common frequency for APRS is 144.39 MHz, although you can use 145.01 and 145.79 MHz. You can find a group of HF APRS users using LSB transmission on 10.151 MHz. The actual tones are below the carrier frequency of 10.151 MHz and fall inside the 30 meter band.

You can do a lot more with APRS than just report location. APRS supports a short-message format similar to the Short Message Service (SMS) format that's used for texting on mobile phones. Popular mapping software offers interfaces so that you can have street-level maps linked to your position in real time by ham radio. Race organizers use APRS to keep track of far-flung competitors. You can also add weather conditions to APRS data to contribute to a real-time automated weather tracking network. To find out more, including detailed instructions on configuring equipment, start with the resources listed in Table 11-9.

Table 11-9 APRS Resources

Resource	*Description*
APRS (`www.aprs.org`)	Website describing the current state of the technology, with useful articles and links
APRS introduction (`www.arrl.org/automatic-packet-reporting-system-aprs`)	Primer on APRS technology and use
OpenAPRS (`www.openaprs.net`)	APRS map server interface
TAPR APRS group (`www.tapr.org/aprs.html`)	Information on APRS equipment, tutorials, mailing lists, and software

Operating via Satellite

Nonhams usually are pretty surprised to find out about ham radio satellites. Imagine — do-it-yourself satellites! The first amateur satellite, OSCAR-1 (Orbiting Satellite Carrying Amateur Radio), was built by American hams and went into orbit in 1961, just a couple of years after the Soviet Union launched Sputnik and ignited the space race. As of early 2013, 25 active satellites were providing ham-to-ham communications or supporting the scientific experiments of student teams by sending telemetry back to Earth.

The main organization for amateur satellite activities is AMSAT (`http://ww2.amsat.org`). AMSAT coordinates the activities of satellite-building teams around the world and publishes the *AMSAT Journal*, which contains some interesting high-tech articles.

Getting grounded in satellite basics

Most amateur satellites are located in near-circular low Earth orbit, circling the planet several a times each day. A few have noncircular Molniya orbits that take them high above the Earth, where they're visible for hours at a time.

Molniya is *lightning* in Russian and is the name given to a Russian fleet of communications satellites that travel in elliptical orbits.

For practical and regulatory reasons, satellite transmissions are restricted to the bands on 15 and 10 meters and to the 2 meter, 70 cm, and microwave bands at 1296 MHz and higher. The ionosphere doesn't pass signals reliably at lower frequencies, and satellite antennas need to be small, requiring shorter wavelengths.

The satellite's input frequencies are called the *uplink,* and the output frequencies are called the *downlink.* The pieces of information that describe a satellite's orbit (and allow software to determine where it is) are called the *orbital* or *Keplerian elements.* Knowing where a particular satellite is in space allows you to operate through it.

There are four common types of satellites:

- **Transponder:** A transponder listens on a range of frequencies on one band, translates those signals to a different band, and retransmits them in real time.
- **Repeater:** Just like terrestrial repeaters, repeater satellites listen and receive on a specific pair of channels. Satellite repeaters are *crossband,* meaning that their input and output frequencies are on different bands.
- **Digital:** Digital satellites can act as bulletin boards or as store-and-forward systems. You can access both types of digital satellites by using regular packet radio protocols and equipment. The International Space Station (ISS) has a digital bulletin board that's available to hams on the ground, as well as an onboard APRS digipeater. Store-and-forward satellites act as message gateways, accepting messages and downloading them to a few control stations around the world. The control stations also pass messages back up to the satellites that are downloaded by ground-based users. Digital satellites are very useful to hams at sea or in remote locations.
- **Telemetry:** Many student teams and other noncommercial groups (whose members have licenses, like all other hams) use amateur radio frequencies to build small satellites called *CubeSats,* which are launched into low Earth orbit as a group when a commercial satellite launch has spare payload capacity. Each CubeSat measures something or performs some interesting function and then sends a stream of digital data *(telemetry)* back to Earth. A CubeSat may or may not be controllable by telecommand from a ham station on Earth. CubeSats typically operate for less than a month; then they gradually reenter the atmosphere and burn up.

If you're interested in supporting or working with a CubeSat team, check out the NASA CubeSat initiative (www.nasa.gov/directorates/heo/home/CubeSats_initiative.html).

Accessing satellites

The best place to find out which satellites are active and in what mode is the AMSAT home page (http://ww2.amsat.org). Click the Sat Status link to get complete information about what each satellite does and its current operational status.

To access satellites, you also need a satellite tracking program. Several of these programs, including free and shareware trackers, are listed at www.dxzone.com/catalog/Software/Satellite_tracking. AMSAT also provides several professional-quality tracking and satellite operation programs.

When you have the tracking software, obtain the Keplerian elements for the satellite you're seeking from the AMSAT home page (click the Keplerian Elements link). Enter this information into your software program, and make sure that your computer's time and date settings are correct.

A complete set of instructions on using satellites is beyond the scope of *Ham Radio For Dummies,* but a short example of how to connect to the packet station aboard the ISS and receive a nice certificate is available at websites like http://cloudpeak.wetpaint.com/page/Packet+and+APRS+on+the+International+Space+Station.

Seeing Things: Image Communication

All the ham transmissions I've covered so far in this chapter have been voice, data, or codes. Don't hams care about pictures and graphics? They do! With the increasing availability of excellent cameras and computer software, getting on one of the amateur image modes has never been easier.

The ease of image communication has resulted in several really interesting uses, such as sending images from balloons and radio-controlled vehicles. In addition, emergency communications teams are starting to use images as tools for assessing damage after a disaster or managing public events. *The ARRL Handbook* and *The ARRL Operating Manual* both have a chapter on image communication. The next few sections discuss these image modes. Figure 11-8 shows examples of images sent on each mode.

Figure 11-8: Pictures typical of those sent via amateur radio image modes.

Courtesy American Radio Relay League

Slow-scan television and facsimile

You can find slow-scan television (SSTV) primarily on the HF bands, where SSB voice transmission is the norm. The name comes from the fact that transmitting the picture over a narrow channel made for voice transmissions takes several seconds. Usually, you can hear slow-scan signals in the vicinity of 14.230 and 21.340 MHz by using USB transmissions.

SSTV enthusiasts start with a webcam or video camera and a sound card. They use frame-grabber software to convert the camera video to data files. Graphics files from any source can be used. SSTV software encodes and decodes the files, which are exchanged as audio transmitted and received with a voice SSB transceiver. You can use analog SSTV, in which the picture is encoded as different audio frequencies, or digital SSTV, in which the picture is broken into individual pixels and transmitted via a digital protocol.

Facsimile over radio is still a widely used method of obtaining weather information from land-based and satellite stations. Hams rarely transmit fax signals anymore, but it's handy to be able to receive fax transmissions.

You can find links to detailed information about SSTV and facsimile transmission at www.ac6v.com/opmodes.htm#SS and www.qsl.net/kb4yz.

Fast-scan television

You can also send full-motion video, just as regular broadcasters do, with fast-scan video transmissions. Fast-scan uses the same video standards as analog broadcast and consumer video, so you can use regular analog video equipment. This mode is usually called *amateur television* (ATV) and is most popular in metropolitan and suburban areas, where transmission distances are relatively short. ATV even has its own repeaters.

ATV transmissions are restricted to the 70 cm band and higher frequencies because of their wide bandwidth — up to 6 MHz. You won't be able to use your regular 70 cm transmitter to handle that bandwidth, so you must construct or purchase a transmitter designed specifically for ATV. The transmitters are designed to accept a regular video camera signal, so little extra equipment except a good antenna is required to use ATV.

Because ATV transmissions use the same video transmission format that analog TV broadcasts did, analog television receivers are used as receivers, with a frequency converter used to transfer the ham band ATV signals to one of the higher UHF TV channels, where they're received just like any other TV signals.

Even though TV broadcasts have gone digital, a lot of analog equipment is still in use, and most digital TVs can display analog video as well. Hams are beginning to use the same types of signals as digital TV broadcasts. As of early 2013, however, suitable equipment is rarely available, and digital ATV remains largely experimental.

Part IV

Building and Operating a Station That Works

You'll really enjoy making custom QSL (contact confirmation) cards for everyday use and special contacts. The Web Extras site at www.dummies.com/extras/hamradio includes a simple model that you can customize and take to a local print shop or office supply store.

In this part . . .

- See how to choose and buy a radio and an antenna.
- Take the next step in setting up your own station safely and effectively.
- Discover QSL cards and the fun process of designing your very own.
- Find out about the "hands-on" part of radio so you can do simple maintenance and connect your equipment properly.
- You'll really enjoy making custom QSL (contact confirmation) cards for everyday use and special contacts. The Web Extras site at `www.dummies.com/extras/hamradio` includes a simple model that you can customize and take to a local print shop or office supply store.

Chapter 12

Getting on the Air

In This Chapter

- Deciding what you want from your station
- Choosing radio equipment
- Picking an antenna
- Selecting a computer
- Buying new or used equipment
- Upgrading your shack

Even a casual stroll through the ads in *CQ* or *QST* turns up page after page of colorful photos, with digits winking, lights blinking, and meter needles jumping. Antennas are even more numerous, with elements sticking out every which way, doodads dripping off them, and all manner of claims made about performance. Then you have to sort through nearly an infinite number of accessories and software packages. The decision can be overwhelming. How do you choose?

This is an exciting time for you and for ham radio. Technology is quickly changing what a radio is and how you interact with it. Radios can now be made completely in software (SDRs). The ever-present Internet gives you the ability to use your radio from anywhere on the planet (remote operating). These developments are just two of the brand-new tools you can use and enjoy, with more becoming available every day at an accelerating pace.

In this chapter, I start by asking questions about what kind of operating you think you'd like to try and what bands you want to use. Then I review the different types of stations and equipment to help you choose among different styles of equipment, select an appropriate antenna, and connect everything. If you want to try the digital modes or just keep a computerized log, this chapter also helps you figure out what kind of computer you need.

Setting Goals for Your Station

Don't tell anybody, but you're about to embark on a journey called system design. You may think that making decisions is impossible, but all you have to do is a little thinking up front.

Deciding what you want to do

You can find many activities in ham radio, which I cover in Part III. You can use the same equipment to participate in most of them. Before you start acquiring equipment, decide what you want to do with it by answering these questions:

- What attracted you to ham radio?
- Can you pick two or three operating activities and modes that really pique your interest?
- If you know and admire a ham, does he or she do something that you want to do?
- Are you most attracted to the shortwave bands or to the VHF and UHF bands?
- What sounds most intriguing: using the digital modes, chatting by voice, or mastering Morse code?
- Will you operate with home, mobile, or portable equipment (or all three)?
- Do you intend to participate mainly for enjoyment or for a specific purpose, such as emergency or travel communications?

All these considerations color your choice of equipment.

Knowing your ham radio resources is also important. Answer these questions:

- What's your budget for getting on the air?
- How much space do you have available for your shack?
- How much space do you have for antennas?

The following sections lay out your options.

Deciding how to operate

The decision to make first is where you expect to operate your station most of the time — at home, in a car, or in a backpack, for example. This choice determines the size and weight of the equipment, what kind of power source it needs, and the type of antennas you'll be using. All those characteristics have a big effect on what features and accessories you'll want and need.

Home operation

A home station is a semipermanent installation. Along with the radio equipment, you need a little furniture and space to put it in. Using the voice modes means speaking out loud and probably listening on a speaker. Using the digital modes or Morse code (CW) is quieter, but you should still put a good pair of comfortable headphones on your shopping list.

Choose a location for your station that minimizes the effect on other family members. A basement shack shouldn't be right below a bedroom, for example. All in all, a spare bedroom or dry basement is about the best place because it won't be wet, hot, or cold.

Because most hams operate with external antennas, plan appropriate ways of getting feed lines to them. What's going to hold the antennas up? Larger structures, such as rotatable beams on masts or towers, may need permits or approvals. (I cover antennas later in this chapter.)

A big part of the amateur service is being available in emergencies. Because you may lose power when you need it most, consider how you might operate your station with the AC power off. A radio that runs on 12 volts can run from a car battery for a while. All your computing gear and accessories also need power. If you have a generator, consider how you can use it to safely power your station, if necessary.

Mobile operation

The small all-band, multiple-mode radios available today enable HF, VHF, and UHF operation from even the smallest vehicle. VHF/UHF antennas are small and inexpensive. Installing an efficient HF mobile antenna is harder, so shift some of your mobile antenna budget toward the HF antenna.

Driving your station creates its own set of unique considerations. Because vehicles come in so many styles, every installation is a custom installation. Leave some budget for automotive fixtures and wiring. You may find it prudent to spend a few dollars for a professional shop to make recommendations about power wiring and safety.

An excellent mobile-station website is Alan Applegate (KØBG)'s Web Site for Mobile Operators, which covers everything from powering your mobile rig to using it effectively.

Portable operation

For purposes of this chapter, assume that *portable* means you carry or pack your entire radio station, including your power source, to the location where you plan to operate. Start by assigning yourself a total weight budget. Get creative with antennas and accessories to maximize your options for the radio and power.

Some amazingly small radios are available. These radios aren't always the easiest to operate, however. If you're just starting, you may want to pass up a minimal radio in favor of one that's easier to operate and has more features until you know more about operating. When you have more experience, you'll know what features you can do without.

As you get started, concentrate on a single band. On HF, the 14, 17, and 21 MHz bands are favorites with low-power and portable operators. These bands are active for a large portion of the day, and the antennas are small enough to carry easily. If you like nighttime operating, 7 MHz and 10 MHz are best.

On VHF, 50 MHz and 144 MHz operation from high spots is common. Plenty of operators are available, particularly during contests, and those bands often feature interesting propagation.

One of the fastest-growing activities is the Summits on the Air (SOTA; `www.sota.org.uk`) award program. It's a favorite with backpackers and hikers who clamber to the top of hills and mountains, put flea-power stations on the air for a little while, and then head back down.

Handheld radio operation

Regardless of what other pursuits you choose in ham radio, you probably want to have a handheld VHF/UHF radio. It's just so darn handy! A handheld radio keeps you in touch with local hams and is very useful on club and personal outings. Many handheld radios also feature an extended receive range that may include commercial broadcast stations, weather stations, or police and fire department bands.

If you're buying your first handheld radio, get a simple, single-band model. You can make a more-informed decision later, when you upgrade to a multiple-band model with all the bells and whistles. Simple radios are also easy to operate. No new radio you can buy today will be missing any significant feature.

Accessories can extend the life and usefulness of a portable radio. Here are some of the most popular:

- The flexible rubber duck antenna supplied with handheld radios is great for portable use but isn't as efficient as a full-size metal antenna. An external base-station antenna greatly extends the range of a handheld radio while you're at home.
- Use a high-quality, low-loss feed line for cables more than a couple dozen feet long. (See "Feed line and connectors," later in this chapter, for more information.)
- A speaker–mike combination allows you to control the radio without having to hold it up to your face.
- A case or jacket (like a smartphone's sleeve) protects the radio against the rough-and-tumble nature of portable use.
- Spare batteries are musts. If you have a rechargeable battery pack, be sure to have a spare, and keep it charged. A drop-in charger works faster than the supplied wall-transformer model. If the manufacturer offers one, a battery pack that accepts ordinary AA cells is good to have, especially in emergencies, when you may not be able to use a charger.

Regardless of what kind of radio you have, be sure to keep a record of model and serial numbers. Engrave your name and call sign in an out-of-the-way location on the case. Mobile and portable radios can be lost or stolen. Even larger radios sometimes get taken outside sometimes. Protect your investment against theft and loss! Check your homeowner's and auto insurance for coverage of radio equipment. If you're an ARRL member, you can use the ARRL equipment insurance program (`www.arrl.org/insurance`).

Allocating resources

When you start assembling a station, you have a range of items to obtain — not only the radio itself, but also antennas, accessories, cables, and power sources. Table 12-1 shows some estimates of relative costs based on the type of station you're setting up. If you pick a radio first, the remaining four columns give you a rough idea of how much you should plan on spending to complete the station. These figures are approximate but can get you started. I assume that all the gear is purchased new.

Table 12-1 Relative Expense Breakdowns

	Radio and Power Supply or Batteries	*Antennas*	*Accessories*	*Cables and Connectors*	*Total Cost Relative to Basic HF Base*
Handheld VHF/UHF	75%	Included	25%	Included	0.3
Mobile VHF/UHF	75%	20%	Not required	5%	0.5
All-mode VHF/UHF	50%	30%	5%	15%	1.0
Portable HF	75%	10%	10%	5%	0.5
Mobile HF	60%	25%	10%	5%	0.7
Basic HF base	50%	25%	15%	10%	1.0
Full-featured HF	75%	15%	10%	5%	2.0

If you prefer digital-mode operating, you need some kind of computer and an interface between it and the radio. This job is perfect for that old laptop! (Also see "Choosing a Computer for the Shack," later in this chapter.)

Choosing a Radio

Now comes the fun part: shopping and choosing!

To get an idea of what products are available, check the advertisements of the latest models in recent copies of *QST and CQ* magazines. If you have a license, no doubt you've received a copy of a catalog from Ham Radio Outlet (`www.hamradio.com`), Amateur Electronic Supply (`www.aesham.com`), or Universal Radio (`www.universal-radio.com`). Perhaps MFJ Enterprises (`www.mfjenterprises.com`) sent you a catalog of an extensive line of accessories. If you have a local radio store, pay a visit to browse through the catalogs and product brochures. Inquire about upcoming sales or promotions. Manufacturers often exhibit their new gear at larger hamfests and conventions. Your job at this point is to gather a wide variety of information.

Dozens of handheld and mobile radios are for sale, so use a checklist of features to help you decide on a model. Note the capabilities you want as well as the ones that fall into the nice-to-have category.

As you can see in Table 12-1 (refer to "Allocating resources," earlier in this chapter), regardless of what kind of station you plan to assemble, a new radio consumes at least half of your budget, which is only appropriate because the radio is the fundamental piece of equipment in ham radio. You interact with the radio more than you do with any other equipment, and a poor performer is hard to compensate for.

So many new hams find this first adventure of buying a radio intimidating that I wrote an article about it for the ARRL. You can download the article at `www.arrl.org/buying-your-first-radio`, along with a checklist for comparisons, other articles about buying and evaluating radios, and a primer on ham jargon.

Radios for the HF bands

All modern radios have perfectly usable receive and transmit performance. The differences involve performance in several key areas, such as capability to receive in the presence of strong-signals, signal filtering and filter control capabilities, coverage of one or more VHF/UHF bands, operating amenities such as subreceivers, and number of built-in antenna tuners.

HF radios for the home station fall into three categories:

- **Basic:** This radio includes a simplified set of controls with basic receiver filters and signal adjustments. Controls may be fixed-value, with on and off settings. The radio has limited displays and metering, connects to a single antenna, and has minimal support for external accessories. A basic radio is good for a beginning ham and makes a great second or portable radio later. A computer control interface may be available.
- **Journeyman:** This radio includes all the necessary receive and transmit adjustments and may have front-panel controls. It has an expanded set of memory, display, and metering functions. You can find models that have additional bands and support for digital data operations. Internal antenna tuners are common, as are connections for external equipment,

such as transverters and band-switching equipment. A computer control interface is included.

- **High-performance:** This radio has an extensive receive and transmit controls on the front panel or controls that are configurable via a menu system. Computer-style screen displays are popular. A state-of-the-art receiver and subreceiver are included, along with complete interfaces for digital data and computer control. Most high-performance radios have *band-scope* or *panadapter* displays that show signals across a wide frequency range. Internal antenna tuners are standard, and some antenna switching usually is provided.

The major ham radio dealers have organized their catalogs and websites to help you find the right radio for what you want to accomplish at a price that fits your budget. Start online, collecting information on the radios you like; then head for the product review sections of eHam.net and QRZ.com. Your club members may have some opinions as well.

ARRL members have access to the detailed product reviews published in *QST*, including complete technical evaluations by the ARRL Lab. The reviews cover everything from top-of-the-line transceivers to microphones and power supplies.

Mobile and portable HF radios

Recognizing the rapid growth in mobile and portable operation, all manufacturers offer small, rugged radios. Each year, more bands and better features are crammed into these amazing radios. These radios are quite capable as base stations if you have limited space at home or desire a dual home/portable station. Many radios include coverage of VHF and UHF bands on the weak-signal modes (SSB and CW) as well as FM.

Be aware that because they're so small, these radios have to make some compromises compared with the high-performance designs. The operator interface is, by necessity, menu-driven. This menu-driven interface makes some adjustments less convenient, although the most-used controls remain on the front panel. The smaller rigs don't include internal antenna tuners at the 100-watt output level, as the larger rigs do.

Where can you fit a radio in your vehicle or boat? If you have an RV or a yacht, you may not have a problem, but in a compact car or an 18-foot runabout, the space issue is quite a challenge. Luckily, many radios designed for mobile use, such as the radio shown in Figure 12-1, have detachable front panels (sometimes called *control heads)*. A detachable panel allows you to put the body of the radio under the dash or a seat or on a bulkhead. If you share the car or boat, get agreement on where to place the radio before drilling any holes.

Figure 12-1: Most mobile radios have detachable front panels, as this radio does.

Courtesy American Radio Relay League

As when evaluating these radios' larger base-station cousins, you have to consider features and accessories. Because these radios have a minimal set of controls and are menu-driven, I recommend that you try one before you buy it. If you don't have a friend who owns one, visit a retail store.

If you can borrow a mobile radio before buying it, make sure that the control head can be mounted in your vehicle where the display is easy to see and the controls are easy to operate. Don't build in a driving distraction! Also make sure that the control head cable will reach the main radio unit, wherever you decide to mount it.

Digital data on HF

More and more HF radios provide a connector or two with a digital data interface built in so that it's easy to connect a personal computer and operate on the digital modes, such as PSK31 or RTTY. A few even have a built-in data modem or a terminal node controller (TNC), which is a type of data modem used for packet radio (see Chapter 11). The key features to look for are accessory sockets on the radio carrying some of the following signals:

- **FSK (frequency-shift keying):** A digital signal at this connector pin causes the transmitter to output the two tones for frequency-shift keying, a method of transmitting using two frequencies, usually used for radioteletype (RTTY).
- **Data In/Out:** If a radio has an internal data modem, you can connect these digital data inputs and outputs to a computer. You may need an RS-232 (a type of serial communication) converter.

- **Line In/Out:** Audio inputs and outputs compatible with the signal levels of a computer's sound card, this input is used for digital data when a computer sound card is used as the data modem.
- **PTT:** This input (the same as the push-to-talk feature on a microphone) allows a computer or other external equipment to turn the transmitter on and off.
- **Discriminator (sometimes labeled DISC):** This input is the unfiltered output of the FM demodulator. External equipment can use this signal both to indicate tuning and to receive data.

To find out how to configure a radio to support digital data, look on the manufacturer's website, or ask the dealer for the radio manual. Proper connections for PSK31 and RTTY operation should be included. If the manual doesn't provide an answer, contact the manufacturer to ask how to hook up the radio.

Searching the web for the combination of your radio model and the digital mode you want to use may turn up interface information for your radio.

HF amplifiers

I recommend that you refrain from obtaining an amplifier for HF operations until you have some experience on the air. Most HF radios output 100 watts or more, which is sufficient to do a lot of operating in any part of the hobby.

You need extra experience to add an amplifier and then deal with the issues of power, feed lines, RF safety, and interference. Also, the stronger signal you put out when using an amplifier affects more hams if the signal is misadjusted or used inappropriately. Perfect the basic techniques first.

When do you need output of 1500 watts, or even 500 to 800 watts? In many circumstances, the extra punch of an amplified signal gets the job done. DXers use amplified signals to make contact over long paths on difficult bands. A net control station's amplifier gets switched on when a band is crowded or noisy so that everybody can hear. In emergencies, an amplifier may get the signal through to a station that has a poor or damaged antenna.

HF amplifiers come in two varieties: vacuum-tube and solid-state. Tubes are well suited to the high power levels involved. Solid-state amplifiers, on the other hand, tend to be complex but require no tuning or warmup; just turn them on and go. Tube amplifiers are less expensive per watt of output power than solid-state amps, but they're larger, and the tubes are more fragile.

Don't attempt to use CB "footlocker"-type amplifiers. These amps are not only illegal, but also often have serious design deficiencies that result in poor signal quality.

VHF and UHF radios

Many HF radios include 50, 144, and 440 MHz operations. The Kenwood 2000 series goes all the way to 1200 MHz. This power makes purchasing a second radio just for VHF/UHF operating less a necessity for the casual operator. Many ham radio shacks have an all-band HF/VHF/UHF radio backed up with a VHF/UHF FM rig for using the local repeaters.

VHF/UHF radios that operate in single sideband (SSB), Morse code (CW), and FM modes are known as *all-mode* or *multimode* to distinguish them from FM-only radios. Many of the VHF/UHF all-mode radios have special features, such as full duplex operation and automatic compensation for transponder offsets, that make using amateur satellites easier. (I introduce amateur satellite operation in Chapter 11.) Satellite operations require special considerations because of the need for cross-band operation and the fact that satellites are moving, which results in a Doppler shift on the received signal.

An all-mode radio can also form the basis for operating on the amateur microwave bands. Commercial radios aren't available for these bands (900 MHz; 2.3, 3.4, 5.6, 10, and 24 GHz; and up), so you can use a transverter instead. A *transverter* converts a signal received on the microwave bands to the 28, 144, or 440 MHz band, where the radio treats it just like any other signal. Similarly, a transverter converts a low-power (100 milliwatts or so) output from the radio back up to the higher band. Bringing the output signal up to 10 watts or more requires an external amplifier.

Multimode radios are very popular for the special type of mobile contesting called *roving.* One of these small rigs and a transverter or two makes for a lot of fun as you rove from grid square to grid square, racking up points on all the bands.

FM-only radios

Nearly every ham uses FM on the VHF and UHF bands regardless of his or her favorite operating style or mode. A newly minted Technician licensee can likely use an FM mobile or handheld radio as his or her first radio. FM is available on the all-mode rigs, but because of the mode's popularity and utility, FM-only rigs are very popular.

FM radios come in two basic styles: mobile handheld and handheld. You can use mobile rigs as base stations at home, too. Both mobile and handheld radios offer a wide set of features, including loads of memory channels to store all your region's repeater information, powerful scanning modes, and several types of squelch systems.

Mobile FM radios

The more-powerful transmitters used with an external antenna extend your range dramatically. Receivers in mobile radios often have better performance than those in handheld radios; they're capable of rejecting the strong signals from commercial transmitters on nearby frequencies. Information about how receivers perform in such conditions is available in product reviews. Your own club members may have valuable experience to share, because they operate in the same places as you.

Although most people expect a new ham to buy a handheld radio, I suggest that you start with a mobile radio shared between your car and at home. Assuming you live in an area with average or better repeater coverage, you can simply pop a magnetic-mount antenna on top of the refrigerator, and you're in business. (If you live in a rural area, you probably need an outdoor antenna.) The stronger signal from the mobile allows you to operate more successfully over a wider range, which is important at first. When you know more about what type of FM operating you want to do, you can buy a handheld radio with the right features and save money.

You can often use mobile radios for digital data operation on the VHF/UHF bands. As radio modem technology has advanced, hams have begun to use 9600-baud data. If you plan to use your mobile rig for digital data, make sure that it's data-ready and rated for 9600 baud without modification.

Handheld FM radios

Handheld radios come in single-band, dual-band, and multiband models. With the multiband radios covering 50–1296 MHz, why choose a lesser model? Expense, for one thing. The single-band models, particularly for 2 meters, cost less than half the price of a multiband model. You'll do the lion's share of your operating on the 2 meter (VHF 144–148 MHz) and 70 cm (UHF 420–440 MHz) bands, so the extra bands may not get much use.

You can expect the radio to include as standard features encoding and decoding of CTCSS subaudible tones (tones used to restrict access to repeaters), variable repeater offsets, at least a dozen memory channels, and a numeric keypad for entering control tones (the same tones used to dial a telephone). A rechargeable battery and simple charger come with the radio. (I discuss repeater operation in Chapter 9.) Be sure to get a spare battery and the base charger, if your budget allows.

Extended-coverage receiving is a useful feature. I find that being able to listen to commercial FM broadcast and weather alert stations around 2 meters is very useful.

Programming a radio with dozens (if not hundreds) of memory channels can be a chore if you do it all with the front-panel controls. Ask around in your club to see whether someone has programming software and a cable to connect your radio to a computer; you'll be really glad you did. The software enables to you to quickly set up your radio for the local repeaters and simplex channels, including alphanumeric labels for each channel, if your radio supports that feature.

Most radios can also be *cloned,* meaning that you can transfer the contents of one radio's memories to another radio of the same model by using a cloning cable. This method can save a lot of time if you buy a new radio and a friend has the same model already programmed.

A rig that's APRS-enabled can connect to (or may include) a GPS receiver and transmits your location as described in Chapter 11.

VHF/UHF amplifiers

Increasing the transmitted power from a handheld or low-power mobile radio is common. Amplifiers can turn a few watts of input into more than 100 watts of output. Solid-state commercial units are known as *bricks* because they're about the size of large bricks, with heat-sinking fins on the top. A small amp and external antenna can improve the performance of a handheld radio to nearly that of a mobile rig.

Amplifiers are either FM-only or SSB/FM models. Amplifiers just for FM use cause severe distortion of a SSB signal. An amplifier designed for SSB use is called a *linear amplifier,* and SSB/FM models have a switch that changes between the modes. You can amplify Morse code signals in either mode, with more gain available in FM mode.

RF safety issues are much more pronounced above 30 MHz because the human body absorbs energy more readily at those frequencies. An amplifier outputs enough power to pose a hazard, particularly if you use a beam antenna. Don't use an amplifier at 50 MHz or above if the antenna is close to people. Revisit your RF safety evaluation if you plan on adding a VHF/UHF amplifier to your mobile or home station.

Software-defined radio

The software-defined radio (SDR) in which most of a radio's functions are implemented by software is one of the most significant changes in ham radio technology since the transistor; I predict that it will largely displace traditional radio construction by 2020. What would happen if the entire radio were implemented in a microprocessor? You can see that day rapidly approaching in FlexRadio's FLEX-6000 series (`www.flexradio.com`). In an SDR, except for the receiver input and transmitter output circuits, the signal-handling part of the "radio" is a program. You can change the program any time you want to make the SDR do something else.

In an SDR, after a some filtering and conditioning, the incoming radio signal is converted to digital data. From there, the signal is all data until it's converted back to audio for the operator — or not, because when you're using the digital modes to communicate, you may not need to hear the signal at all; you need only to see characters on the screen. Going the other way, the outbound signal only needs to be amplified before being applied to the antenna and radiated into space.

Thus far, several SDR ham radio transceivers are available, and hams are finding the latest models to be strong performers. (For receiving only, SDRs have even been built into USB memory-stick form factors.) The Elecraft KX3 contains all the necessary hardware in a stand-alone box; it looks more or less like a traditional radio, although the controls have some different functions. The FlexRadio 6000-series of SDRs uses a PC as the control and user interface, as shown in Figure 12-2. On a PC, a typical SDR displays signals across a frequency range as in the top of the figure. This display is called a *panadapter* view. Below it is the *waterfall* display, with colors that show which signals are strongest over time. You use a mouse to select signals and click operating controls.

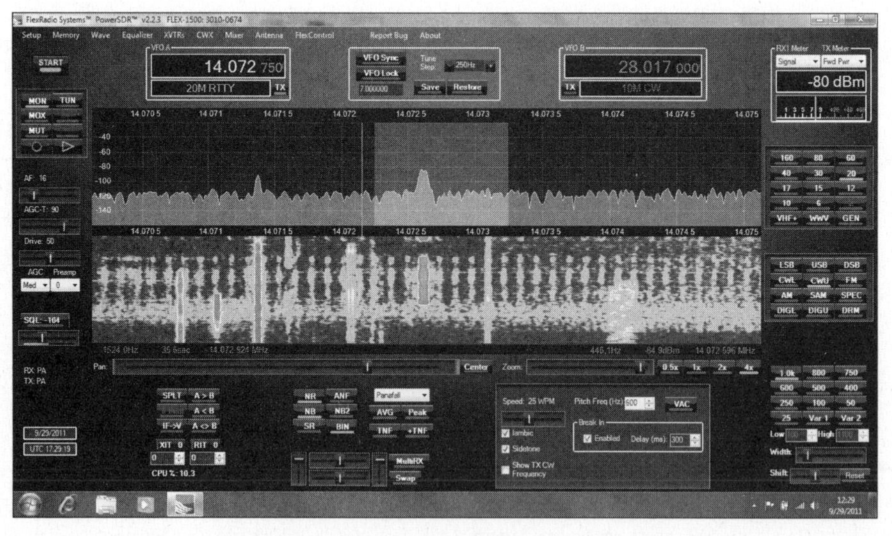

Figure 12-2: The PowerSDR software running with a FlexRadio.

Hams are just beginning to scratch the surface of what SDR can do. All that data onscreen can just as easily be passed to more programs. What will hams do with it? Nobody knows yet. It's an exciting time for ham radio!

Choosing Filters

After picking up a weak signal, the most important capability for a receiver is distinguishing one signal from another. This capability is called *selectivity.* Receivers use filters to remove unwanted signals from interfering with the one you want to receive while *attenuating* (reducing the strength of) unwanted signals just a few hundred hertz away. For many years, the only components that could accomplish this feat were quartz crystals making up a *crystal filter,* which is the type of filter you'll find in most used radios. Nearly every crystal filter has a fixed *bandwidth* (the range of frequencies it can pass, measured in hertz) and can't be adjusted.

Fixed-width filters are available with several bandwidths. A radio is always shipped with at least one SSB filter (HF radios) or FM filter (VHF handhelds and mobiles) installed. The standard filter bandwidth for HF SSB operation is 2.4 to 2.8 kHz. Filters with widths of 1.5 to 2.0 kHz are available for operating under crowded conditions with some loss of fidelity. For Morse code and digital modes, the standard filter is 500 Hz wide and is a good option to select. Narrower filters, down to 250 Hz, are available. The most common filter option is the 500 Hz CW filter, followed by a narrow SSB filter.

Mid-level and high-end radios manufactured today, however, usually use DSP technology (see the nearby sidebar) to do the same thing in software. DSP performs as well as fixed-width filters, and the bandwidth of DSP filters usually is adjustable, which allows you to use just the right amount of filtering. DSP can also perform filtering functions to remove off-frequency signals, reduce or eliminate several kinds of noise, or automatically detect and remove interfering tones. Radios with DSP often use fixed-width *roofing filters* that help the receiver reject extremely strong signals often encountered on busy bands or near shortwave broadcast stations. If roofing filters are available for your radio, purchasing one with a bandwidth of a few kHz is a good way to get the best performance from the receiver.

Digital signal processing

Digital signal processing (DSP) refers to a microprocessor in the radio running special software that operates on, or *processes,* incoming signals, usually on the audio signal before it's amplified for output to the speaker or headphones. More advanced DSP can act on the signals at radio frequencies. In general, the higher the number of bits specified for DSP and the higher the frequency at which the DSP functions are performed, the better the DSP processing performs. (Look in the radio's operating manual or product specification sheet for more information.)

Choosing an Antenna

I can't say which is more important: the radio or the antenna. Making up for deficiencies in one by improving the other is difficult. A good antenna can make a weak radio sound better than the other way around. You need to give antenna selection at least as much consideration you do to radio selection. This section touches on several types of useful and popular antennas.

If you want to know more about antennas and want to try building a few yourself, you need more information. I can think of no better source than *The ARRL Antenna Book,* now in its 22nd edition. It's a good ham resource, and many professional antenna designers have a copy, too. Also, I include a list of useful antenna design books and websites in Appendix B. If you need an introduction to antennas, try the ARRL book *Basic Antennas.*

VHF/UHF antennas

Most antennas used above 50 MHz at fixed stations are either short *whips* (thin steel or aluminum rods) or beams. Whips mounted so that they're vertical are used for local FM operation, whereas beams — antennas that transmit and receive in a preferred direction — are used for VHF/UHF DXing (contacting a distant station) on SSB and CW. The most common type of beam antenna is called a *Yagi,* after Japanese scientists Yagi and Uda, who invented the antenna in the 1920s. A Yagi has several straight rods or tubes (called *elements)* mounted on a long supporting tube called a *boom. Log-periodics* that look like large TV antennas with lots of elements are also types of beam antennas used by hams.

Polarization — the orientation with respect to the ground of the antenna and the radio waves from it — is most important on the VHF and UHF bands, where signals usually arrive with their polarization largely intact. If the radio wave and the antenna are oriented differently, the antenna won't receive the radio wave very effectively. On the HF bands, travel through the ionosphere scrambles the polarization so that it's much less important.

FM operating is done with vertically polarized antennas because vertical antennas on vehicles are much simpler to construct and install. Vertical antennas are also *omnidirectional,* meaning that they transmit and receive equally well in all directions. These characteristics are important for mobile operation — the first widespread use of FM. To prevent cross-polarization, the base antennas are vertical. This convention is universal.

A popular and inexpensive vertical antenna is the simple *quarter-wave whip,* or *ground-plane,* antenna. Many hams build a short ground-plane antenna as a first antenna project.

Operators chasing long-distance VHF and UHF contacts use beam antennas that are horizontally polarized. Many of the long-distance VHF and UHF propagation mechanisms respond best to horizontally polarized waves. If you have an all-mode radio and want to use it for both FM and SSB/CW/digital operating, you'll need both vertically and horizontally polarized antennas.

If you choose to use a beam on VHF and UHF bands, it's a good idea to use 3 to 5 elements on 6 meters and 5 to 12 elements at higher frequencies. These antennas are small enough to mount and turn with heavy-duty TV antenna hardware.

HF antennas

At HF, antennas can be fairly large. An effective antenna is usually at least ¼ wavelength in some dimension. On 40 meters, for example, a ¼-wavelength vertical antenna is a metal tube or wire 33 feet high. At the higher HF frequencies, antenna sizes drop to 8–16 feet but are still larger than even a big TV antenna. Your physical circumstances have a great effect on what antenna you can put up. Rest assured that a large variety of designs can get you on the air.

Wires, verticals, and beams are the three basic HF antennas used by hams all over the world. You can build all these antennas with common tools or purchase them from the many ham radio equipment vendors.

Wire antennas

The simplest wire antenna is a *dipole,* which is a piece of wire cut in the middle and attached to a feed line, as shown in Figure 12-3. The dipole gives much better performance than you may expect from such a simple antenna. To construct a dipole, use 10- to 18-gauge copper wire. It can be stranded or solid, bare or insulated.

While you'll often see the formula for dipole length given as 468/f, the building process is a lot more reliable if you begin with a bit longer antenna and make one or two tuning adjustments. Start with an antenna length of

Length in feet = 490 / frequency of use in MHz

Allow an extra 6 inches on each end for attaching to the end insulators and tuning and another 12 inches (6 inches × 2) for attaching to the center insulator. If you're building a 10 meter dipole, you should start with a length of 490 / 28.3 = 17.3' + 6" + 6" + 12" = 19.3 feet.

To assemble the dipole, follow these steps:

1. **Cut the wire exactly in the middle, and attach one piece to each end insulator, just twisting it back on itself for the initial check.**
2. **Attach the other end to the center insulator in the same way.**
3. **Attach the feed line at the center insulator, and solder each connection.**
4. **Attach some ropes, and hoist the antenna in the air.**
5. **Check the dipole.**

   Make some short, low-power transmissions to measure the standing wave ratio (SWR, a measure of RF energy reflected back to the transmitter by the antenna), as explained in your radio's operating manual, or use an antenna analyzer such as the MFJ-259B (`www.mjfenterprises.com`). The SWR should be somewhat less than 2:1 on the frequencies you want to use.
6. **Adjust the antenna's length to SWR.**

   If the SWR is low enough but at too high a frequency or is lowest at the high end of the band, loosen the connections at the end insulators and lengthen the antenna by a few inches on each end.

   If the frequency of lowest SWR is too low, shorten the antenna by the same amount.
7. **After you adjust the antenna length so that the SWR is satisfactory, make a secure wrap of the wire at the end insulators, solder the twist if you like, and trim the excess.**

   You've made a dipole!

You can follow the same steps for most simple wire antennas; just vary the lengths.

Figure 12-3 shows some examples of simple wire antennas you can build and install in your backyard or use for portable operation. Most cost only a few dollars and can be built and installed in an afternoon. There's a special sense of satisfaction in making contacts via antennas that you built yourself.

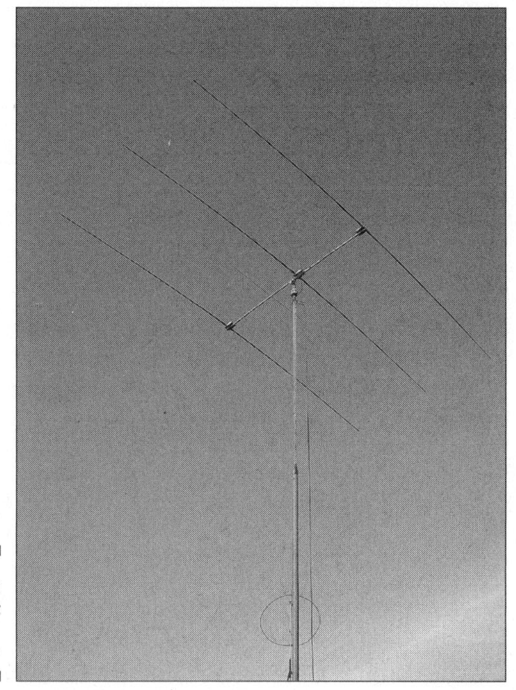

Figure 12-3: Simple HF antennas.

You can connect the dipole directly to the transmitter with coaxial cable and use the dipole on the band at which it's ¼ wavelength long or any odd number of ½ wavelengths. A dipole tuned for 7 MHz, for example, works reasonably well on the 21 MHz band, too, where it's approximately 3½ wavelengths long.

Other common and simple wire antenna designs include

- **Inverted-V:** This dipole is supported at its midpoint, with the ends angling down at up to 45 degrees. This antenna requires only one tall, central support and gives good results in nearly all directions.

- **Full-wavelength loop:** Attach a feed line at the middle of the loop's bottom and erect the loop so that it's vertical. Then the feed line works best broadside to the plane of the loop. These antennas are larger than dipoles but radiate a little more signal in their favored directions.
- **Multiple-band dipole:** The wires of this antenna are fed at the center with *open-wire* or *ladder-line* feed line and used with an antenna tuner to cover several bands. These antennas usually aren't ½ wavelength long on any band, so they're called *doublets* to distinguish them from the ½-wavelength dipoles.
- **Trap dipole:** This antenna uses some appropriately placed components to isolate portions of the antenna at different frequencies so that the dipole acts like a simple ½-wavelength dipole on two or more bands.
- **Random-length wire:** Attach some open-wire feed line 15 to 35 feet from one end, and extend the wire as far and high as you can. A couple of bends won't hurt. You have to convert the feed line to coaxial cable using a balun or antenna tuner. This antenna's performance is hard to predict, but it's an excellent backup or temporary antenna.

For more information on these and many other antennas you can build, check out some of the references listed in Appendix B.

If you don't have the perfect backyard to construct the antenna of your dreams, don't be afraid to experiment. Get an antenna tuner (or use the one in your radio), and put up whatever you can. You can even bend wires or arrange them at strange angles. Antennas want to work!

Vertical antennas

Vertical antennas are nearly as popular as wire antennas. The ¼-wavelength and ½-wavelength antennas are two common designs. Verticals don't require tall supports, keep a low visual profile, and are easy to move or carry. Verticals radiate fairly equally in all horizontal directions, so they're considered to be *omnidirectional* antennas.

The ¼-wavelength design is a lot like a ½-wavelength dipole cut in half and turned on end. The missing part of the dipole is supplied by an electrical mirror of sorts, called a *ground screen* or *ground plane.* A ground screen is made up of a dozen or more wires stretched out radially from the base of the antenna and laid on top of the ground. The feed line connects to the vertical tube (it can also be a wire) and to the radials, which are all connected. The ¼-wavelength verticals are fairly easy to construct and, like dipoles, work on odd multiples of their lowest design frequency.

Ground-independent verticals are about twice as long as their ¼-wavelength counterparts but don't require a ground screen. The lack of a ground screen means that you can mount them on masts or structures above the ground. The feed line is connected to the end of the vertical but requires a special impedance matching circuit to work with low-impedance coaxial feed lines. Several commercial manufacturers offer ground-independent verticals, and many hams with limited space or opportunities for traditional antennas make good use of them.

Both types of verticals can work on several bands through the use of the similar techniques used in wire antennas. Commercial multiple-band verticals work on up to nine of the HF bands.

Beam antennas

The most common HF Yagi today is a three-element design (a reflector, a driven element, and a director) that works on three popular ham bands (20, 15, and 10 meters) and so is called a *tri-bander.* Figure 12-4 shows a three-element Yagi beam on a 55-foot mast whose lowest operating frequency is 14 MHz.

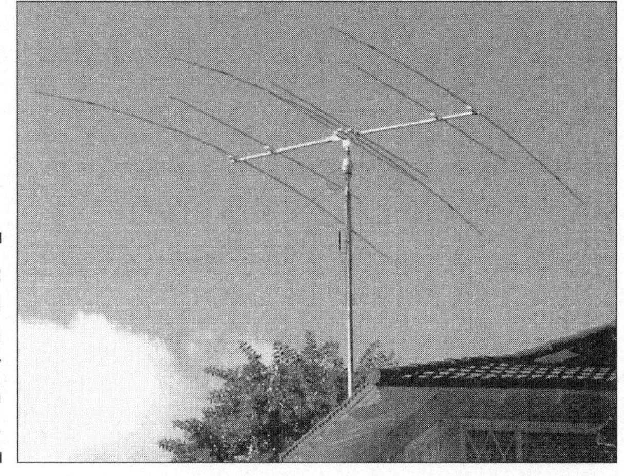

Figure 12-4: A typical HF beam antenna for 20 through 10 meters.

Other HF beams are made from square or triangular loops. They work on the same principle as the Yagi, but with loops of wire instead of straight elements made from rod or tubing. Square-loop beams are called *quads,* and the triangles are called *delta loops.* Log periodics are also used on the HF bands, with popular models available that cover 20 through 10 meters.

Whereas wire antennas have a fixed orientation and verticals radiate equally in all horizontal directions, a beam antenna such as the one shown in Figure 12-4 can be rotated, which allows you to concentrate your signal or reject an interfering signal in a certain direction. You can place small HF beams on inexpensive masts or rooftop tripods, although they overload most structures designed for TV antennas. You also need a *rotator* that mounts on the fixed support and turns the beam. You can control the rotator from inside the shack with a meter to indicate direction (see "Supporting Your Antenna," later in this chapter).

Most hams start on HF with a wire or vertical antenna. After you operate for a while, the signals you hear on the air give you a good idea of what antennas are effective. After you have some on-the-air experience, you can decide whether you need a beam antenna.

Mobile and portable antennas

For VHF/UHF FM use, mobile antennas are whips that are at least a full ¼ wavelength.

You can mount mobile antennas as removable or permanent fixtures. The most easily removed antennas are the *mag-mount* models, which use a magnetic base to hold themselves to a metal surface. Mag-mount antennas are available from HF through UHF. The drawback is that the installation isn't as clean as that of a permanently mounted antenna. Trunk-mount antennas for VHF and UHF are semipermanent and look better. Drilling a hole in your car for a permanent mount looks best of all. All three options are fairly close in performance. Whichever method you choose, be sure that you can remove the antenna from the mount to deter theft and for clearance, such as in a car wash or parking garage. You can generally route antenna cables under trim, carpet, and seats.

For mobile operation on the HF bands, many hams use Hamstick-type antennas attached to a mag-mount or a permanent mount attached directly to the vehicle, as shown in Figure 12-5. These antennas are wire coiled on fiberglass tubes about 4 feet long, with a stainless-steel whip attached at the top. The antennas work on a single band and are sufficiently inexpensive that you can carry a whole set in the car. They require you to change the antenna when changing bands. Another design uses resonators attached to a permanent base section to operate on different bands. The resonators and fiberglass whip antennas use a standard ⅜-24 threaded mount.

An adjustable design that has become popular in recent years has a moving top section that allows the antenna to tune over nearly any HF frequency. Antennas of this type are known as *screwdriver* antennas because they use DC motors similar to those in battery-powered electric screwdrivers. Hamstick-type antennas are least expensive, and screwdriver antennas are most expensive. Performance varies dramatically, depending on mounting and installation, so guaranteeing good results for any of the styles is difficult.

Good results from a mobile station are much easier to obtain on the 20 meter and higher frequency bands. The need to keep antennas small makes operating on the "low bands" (40 meters and below) more of a challenge.

At VHF and UHF, portable antennas are very small, light, and easy to pack and carry. At HF, however, the larger antennas are more difficult to deal with. You can try a lightweight wire antenna if you can find a way to support it well above the ground. Trees or lightweight fiberglass masts are your best choices. Vertical antennas need a sturdy base and usually a set of wires to make a ground screen. Telescoping antennas may be options, and you can use the mobile Hamstick-type whips with a few radial wires.

An antenna that doesn't present a low SWR requires a tuner for the transmitter to output full power, adding weight and expense. The smaller coaxial feed lines, such as RG-174 and RG-58, also have higher losses. Try out the performance of your antenna and feed line before taking off on a major adventure to avoid unpleasant surprises.

Figure 12-5: A Hamstick-type mobile antenna mounted on the NØAX mobile along with a small VHF/UHF dual-band whip.

Answering the decibel

Time for a refresher on decibels? Decibels (dB) are used to measure a ratio of two quantities in terms of factors of 10. A change of a factor of 10 (from 10 to 100 or from 1 to 0.1) is a change of 10 dB. Increases are positive, and decreases are negative. You can use the following formula to calculate dB for changes in power and voltage:

```
dB = 10 log (power ratio)
dB = 20 log (voltage ratio)
```

Decibels add if ratios are multiplied together. Two doublings of power `(x 2 x 2)`, for example, is `3 dB + 3 dB = 6 dB`. A gain of 20 can also be expressed as `x 10 x 2 = 10 dB + 3 dB = 13 dB`.

If you memorize these dB-ratio pairs, you can save yourself a lot of calculating because you won't need a precise dB calculation very often:

Power x 2 = 3 dB	Power x 1/2 = -3 dB
Power x 4 = 6 dB	Power x 1/4 = -6 dB
Power x 5 = 7 dB	Power x 1/5 = -7 dB
Power x 8 = 9 dB	Power x 1/8 = - 9 dB
Power x 10 = 10 dB	Power x 1/10 = -10 dB
Power x 20 = 13 dB	Power x 1/20 = -13 dB
Power x 50 = 17 dB	Power x 1/50 = -17 dB
Power x 100 = 20 dB	Power x 1/100 = -20 dB

A change of 1 receiver S unit represents approximately 6 dB.

Feed line and connectors

Gee, how tough can picking a feed line be? It's just coax, right? Not really, and you may be surprised by how much feed line can affect your signal, both on transmit and receive. When I started hamming, I used 100 feet of RG-58 with a 66-foot dipole that I tuned on all bands. I didn't know that on the higher bands, I was losing more than half of my transmitter output and received signals in the coax. The 50-foot piece I was using on my 2 meter antenna lost an even higher fraction of the signals.

Although losing 50 percent, or 3 dB (decibels), is only about half an S unit (S units measure signal strength; a change of 1 S unit represents a factor of 4 times higher or lower power), I lost a few contacts when signals were weak or the band was noisy. Making up that loss with an amplifier costs hundreds of dollars. Changing antennas to a beam with 3 dB of gain costs at least that much, not counting the mast and rotator. That makes the extra $20 to $25 cost of lower-loss RG-213 cable look like a pretty good bargain.

Table 12-2 compares several popular feed lines in terms of their relative cost (based on RG-58) and the loss for a 100-foot section at 30 MHz and 150 MHz. The loss is shown in dB and in S units on a typical receiver, assuming that one S unit is equivalent to 6 dB.

Table 12-2 Relative Cost and Loss of Popular Feed Lines

Type of Line and Characteristic Impedance	*Outside Diameter (Inches)*	*Cost Per Foot Relative to RG-58*	*Loss of 100' at 30 MHz in dB and S Units*	*Loss of 100' at 150 MHz in dB and S Units*
RG-174A/U (50 ohms)	0.100	⅔	6.4 dB/1 S unit	>12 dB/>2 S units
RG-58C/U (50 ohms)	0.195	1	2.6 / 0.5	6.7 / 1.1
RG-8X (50 ohms)	0.242	1¼	2.0 / 0.3	4.6 / 0.7
RG-213/U (50 ohms)	0.405	3	1.2 / 0.2	3.1 / 0.5
1" ladder line	½" or 1" width	1 to 2	0.1 / <0.1	0.4 / <0.1

The moral of the story is to use the feed line with the lowest loss you can afford. Open-wire feed line is a special case because you add an impedance transformer or tuner to present a 50-ohm load to the transmitter, incurring extra expense and adding some loss.

To save a lot of money on feed line, buy it in 500-foot spools from a distributor. If you can't afford to buy the entire spool, share the spool with a friend or two. Splitting the expense is an excellent club buy and can save more than 50 percent compared with buying cable 50 or 100 feet at a time. Do the same for coaxial connectors.

Beware of used cable unless the seller is completely trustworthy. Old cable isn't always bad but can be lossy if water has gotten in at the end or from cracks or splits in the cable jacket. If the cable is sharply bent for a long period, the center conductor can migrate through the insulation to develop a short or change the cable properties. (Migration is a particular problem with foam-insulation cables.)

Before buying used cable, examine the cable closely. The jacket should be smooth and shiny, with no obvious nicks, dents, scrapes, cracks, or deposits of adhesive or tar (from being on a roof or outside a building). Slit the jacket at each end for about 1 foot and inspect the braid, which should be shiny and show no signs of corrosion or discoloration whatsoever. Slip the braid back. The center insulator should be clean and clear (if solid) or white if foam or Teflon synthetic. If the cable has a connector on the end, checking the cable condition may be difficult. Unless the connector is newly installed, you should replace it anyway, so ask if you can cut the connector off to check the cable. If you can't cut it off, you probably shouldn't take a chance on the cable.

The standard RF connectors used by hams are BNC, UHF, and N-type connectors. BNC connectors are used for low power (up to 100 watts) at frequencies through 440 MHz. UHF connectors are used up to 2 meters and can handle full legal power. N connectors are used up through 1200 MHz, can handle full legal power, and are waterproof when properly installed. (Photos or drawings of connectors are shown in vendor catalogs, such as The RF Connection catalog at `www.therfc.com`.)

Good-quality connectors are available at low prices, so don't scrimp on these important components. A cheap connector works loose, lets water seep in, physically breaks, or corrodes, eating up your valuable signals. By far the most common connector you'll work with is the PL-259, the plug that goes on the end of coaxial cables. The Amphenol 83-1SP model is the standard PL-259 connector. By buying in quantity, you can get these high-quality connectors at a steep discount compared with purchasing them individually.

Installing PL-259 connectors is part of ham radio's technical side and requires a bit of craft and skill. Ask your Elmer to show you how to do it properly, with all the connections soldered and then waterproofed.

To crimp or not to crimp? You can install a crimp-on connector quickly and get reliable service if you do it right. Crimped connectors also work outside but absolutely require proper waterproofing, which means more than a layer of electrical tape. Crimping tools, or *crimpers,* are available for less than $50. If you have a lot of connections to make, a crimp-on connector may be a good choice.

Waterproofing connectors requires attention to detail and the proper materials. Water in connectors causes many problems at radio frequencies, so learn how to do it right every time, Waterproofing techniques for the common PL-259 are listed in the *ARRL Handbook* and the *ARRL Antenna Book*.

Supporting Your Antenna

Antennas come in all shapes and sizes, from the size of a finger to behemoths that weigh hundreds of pounds. All antennas, however, need to be clear of obstacles and kept away from ground level, which is where most obstacles are.

Before you start mounting your antenna, take a minute to review some elementary safety information for working with antennas and their supports. The article "Antenna and Tower Safety" at www.arrl.org/antennas points out a few common pitfalls in raising masts and towers. To work with trees, aside from using common sense about climbing, you may want to consult an arborist.

Antennas and trees

Although Marconi used a kite for his early experiments, a handy tree is probably the oldest antenna support. A tree often holds up wire antennas, which tend to be horizontal or use horizontal support ropes. The larger rotatable antennas and masts are rarely installed on trees (not even on tall, straight conifers) because of the mechanical complexity, likelihood of damage to the tree, and mechanical interference between the antenna and tree.

Nevertheless, for the right kind of antennas, a tree is sturdy, nice to look at, and free. The goal is to get a pulley or halyard into the tree at the maximum height. If you're a climber (or can find someone to climb for you), you can place the pulley by hand. Otherwise, you have to figure out some way of getting a line through the tree so you can haul up a pulley. You may be able just to throw the antenna support line over a branch. Bear in mind that when a line is pressing against the bark of a tree, the tree can rub and chafe against the line until the line breaks. (This catastrophe always happens at night, in a storm, or right before an important contact.) If the line stays intact, the tree tries to grow around the line, creating a wound that makes raising or lowering the antenna impossible.

If you intend for the tree to support an antenna permanently, bringing in an arborist or a tree service professional to do the job right, using sturdy, adequately rated materials, is worth the expense. Radio Works has a good introduction to antennas in trees at www.radioworks.com.

Masts and tripods

A wooden or metal mast is an inexpensive way to support an antenna up to 30 feet above ground. If you're handy with tools, making a homebrew mast is a good project; numerous articles about their construction are in ham magazines. Masts are good candidates to hold up wire HF antennas and VHF/UHF antennas, such as verticals and small beams. If you're just supporting a VHF or UHF vertical, you won't need a heavy support and probably can make a self-supporting mast that doesn't need guy wires. If you area has high winds or if the mast is subjected to a side load (such as for a wire antenna), however, it needs to be guyed.

One commercially available option is a telescoping *push-up mast,* designed to hold small TV antennas and often installed on rooftops. Push-ups come in sizes up to 40 feet, with guying points attached. You can also construct masts by stacking short sections of metal TV antenna mast, but you have to add your own guying points. You can't climb either telescoping or sectional masts, so mounting the antenna and then erecting the whole assembly is up to you. You can also mount a section or two of stacking mast on a chimney to support a small vertical. Push-up and TV masts are available (along with all the necessary mounting and guying materials) at RadioShack and at hardware and home-improvement stores.

One step beyond the mast is the roof-mount tripod. The lighter tripods are used for TV antennas and can hold small amateur antennas. Larger tripods can handle midsize HF beams. Tripods are good solutions in urban areas and in subdivisions that may not allow ground-mounted towers. Tripods are available from several tower and antenna manufacturers.

In recent years, multiple-section telescoping fiberglass masts have become available at low cost, such as those from Spiderbeam (`www.spiderbeam.us`). These masts can't hold a lot of weight but are perfect for small wire antennas and verticals. They make great portable antenna supports, too.

Towers

By far the sturdiest antenna support is the tower. Towers are available as self-supporting (unguyed), multisection crank-up, tilt-over, and guyed structures 30 feet tall and higher. Towers are capable of handling the largest antennas at the highest heights, but they're substantial construction projects, usually requiring a permit to erect. Table 12-3 lists several manufacturers of towers.

Table 12-3 Tower Manufacturers

Antenna	*Website*	*Lattice*	*Crank-Up*	*Self-Supporting*
U.S. Towers	Sold through Ham Radio Outlet (`www.hamradio.com`)		X	
Rohn Industries*	`www.rohnnet.com/Index.htm`	X		X
Trylon	`www.trylon.com`			X
Heights Tower	`www.heightstowers.com`	X	X	X
Universal Aluminum Towers	Sold through distributors			X

** Rohn was purchased by Radian, a Canadian company that plans to continue to manufacture tower components to the Rohn specifications. As of mid-2013, it wasn't clear whether the Rohn name would be used.*

The most common ham tower is a *welded lattice tower.* This tower is built from 10-foot sections of steel tubing and welded braces, and you must guy or tie it to a supporting structure, such as a house, at heights of 30 feet or more. A modest concrete base of several cubic feet is required to provide a footing. Lattice towers for amateur use are 12 to 24 inches on a side and can be used to construct towers well over 100 feet high. Lattice towers are sufficiently strong to hold several large HF beam antennas, if properly guyed. *Tilt-over towers* are lattice towers hinged in the middle so that you can pivot the top sections toward the ground by using a winch. Because of mechanical considerations, tilt-overs are limited to less than 100 feet in height.

Crank-up towers are constructed from telescoping tubing or lattice sections. A hand-operated or motorized winch raises and lowers the tower with a cable and pulley arrangement. A fully nested crank-up is usually 20 to 25 feet high, reducing visual effect on the neighborhood, and when fully nested and blocked for safety you can climb it to work on the antennas. Crank-ups also usually have a tilting base that aids in transporting and erecting the tower. Crank-ups are unguyed, so they depend on a massive concrete foundation of several cubic yards to keep their center of gravity below ground level to prevent tipping over when fully extended. You can install crank-ups in small areas where guying isn't possible; they're available in heights of up to 90 feet.

Self-supporting towers are triangular in cross section, are constructed of trusslike sections like lattice towers, and rely on a large concrete base for center-of-gravity control. They're simpler and less expensive than crank-ups. Available at up to 100 feet, they have carrying capacity similar to that of lattice towers. Mounting antennas along the length of a self-supporting tower is more difficult than for a lattice tower with vertical supporting legs.

Be extremely careful when buying a used tower or mast. Unless the item has been in storage, exposure to the elements can cause corrosion, weakening welds and supporting members. If it's disassembled improperly, the tower can be damaged in subtle ways that are difficult to detect in separate tower sections. A tower or mast that has fallen is often warped, cracked, or otherwise unsafe. Have an expert accompany you to evaluate the material before you buy.

You can construct self-supporting towers from unorthodox materials such as telephone poles, light standards, well casing, and so on. If a structure of any sort can hold up an antenna, rest assured that a ham has used one to do so at some time. The challenge is to transport and erect the mast, climb it safely, and create a sturdy antenna mounting structure at the top.

Regardless of what you decide to use to hold up your antennas, hams have a wealth of experience to share in forums such as the TowerTalk e-mail list. You can sign up for TowerTalk at `www.contesting.com`. The topics discussed range from mounting verticals on a rooftop to which rope is best to giant HF beams and how to locate true north. The list's members include experts who can handle some of the most difficult questions.

When you join an online community like TowerTalk, before you start asking questions, check the archived messages and any FAQ files or other documents that may be available. The longtime members will appreciate it.

Rotors or rotators

A *rotor* is a part of a helicopter and has nothing whatever to do with ham radio. A *rotator,* on the other hand, is a motorized gadget that sits on a tower or mast and points antennas in different directions. Rotators are rated in terms of wind load, which is measured in the number square feet of antenna surface it can control in strong winds. If you decide on a rotatable antenna, you need to figure out its wind load to determine the size rotator it requires. Wind load ratings for antennas are available from antenna manufacturers. The most popular rotators are made by Hy-Gain (`www.hy-gain.com`) and Yaesu (`www.yaesu.com`).

You need to be sure that you can mount the rotator on your tower or mast; some structures may need an adapter. Antennas mount on a pipe mast that then sits in a mast clamp on top of the rotator. If you mount the rotator on a mast, as shown on the left side of Figure 12-6, you must mount the antenna right at the top of the rotator to minimize side loading. This type of installation reduces the maximum allowable antenna wind loading by about half.

In a tower, the rotator is attached to a *rotator plate* (a shelf inside the tower for the rotator to sit on), and the mast extends through a *sleeve* or *thrust bearing* (a tube or collar that hold the mast centered above the rotator), as shown in the center and on the right side of Figure 12-6. Because using a bearing in this way prevents any side loading of the rotator, you can mount antennas well above the tower top if the mast is sufficiently strong.

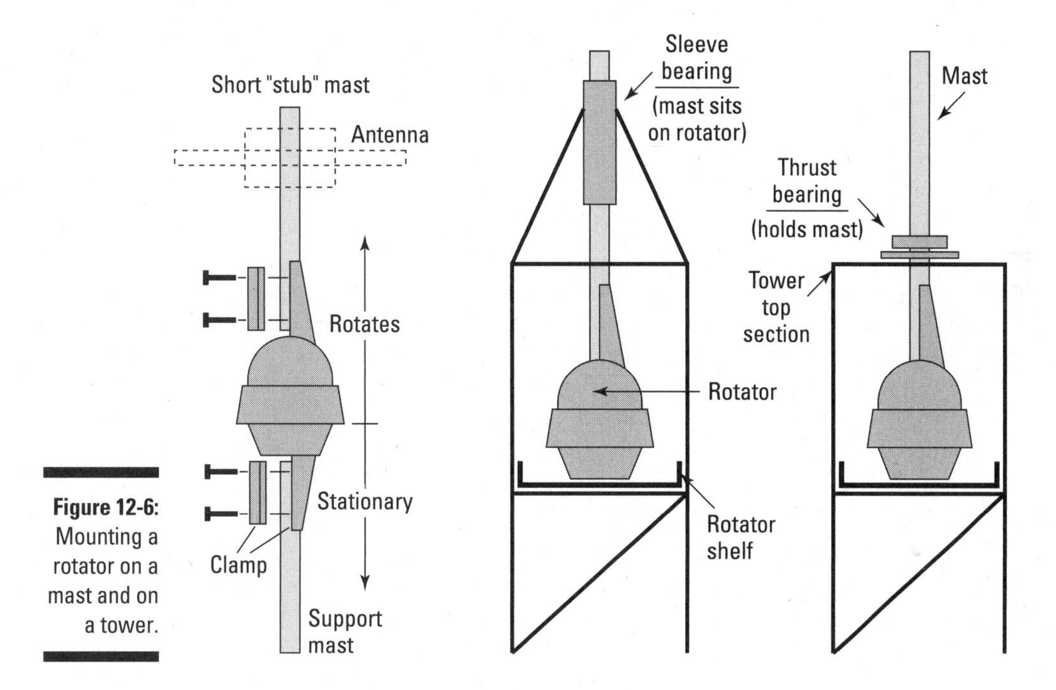

Figure 12-6: Mounting a rotator on a mast and on a tower.

An indicator assembly called a *control box,* which you install in the shack, controls the rotator. The connection to the rotator is made with a multiple-conductor cable. Install the feed lines to the antennas in such a way so that they can accommodate the rotation of the antennas and mast.

Used rotators can be risky purchases. They're always installed in exposed locations and wear out in ways that aren't visible externally. Even if the rotator turns properly on the bench, it may jam, stall, or slip under a heavy load. Buy a new rotator or get a used one from a trusted source for your first installation.

Radio accessories

You can buy or build hundreds of gadgets to enhance whatever style or specialty you choose. Here's some information on the most common accessories that you need to get the most out of your station.

Mikes, keys, and keyers

Most radios come with a hand microphone, although if you buy a used radio, the hand mike may be long gone or somewhat worn. The manufacturer-supplied hand mikes are pretty good and are all you need to get started. After you operate for a while, you may decide to upgrade.

If you're a ragchewer, some microphones are designed for audio fidelity with a wide frequency response. Net operators and contesters like the hands-free convenience of a headset with an attached boom mike held in front of your mouth. Handheld radios are more convenient to use, with a speaker–microphone combination accessory that plugs into your radio and clips to a shirt pocket or collar. Your radio manufacturer may also offer a premium microphone as an option or accessory for your radio. Heil Sound (`www.heilsound.com`) manufactures a wide range of top-quality microphones and headsets.

The frequency response of a microphone can make a big difference on the air. If you operate under crowded conditions, the audio from a microphone whose response emphasizes the midrange and higher frequencies is more likely to cut through the noise. Some microphones have selectable frequency responses so that you can have a natural-sounding voice during a casual contact and then switch to the brighter response for some DXing. If you're not sure which is best, ask the folks you contact, or do an over-the-air check with a friend who knows your voice.

Morse code enthusiasts have thousands of keys to choose among, spanning more than a century of history (see Chapter 11). Beginners often start with a straight key and then graduate to an electronic keyer and a paddle. If you think you'll use CW a lot, I recommend going the keyer/paddle route right away. Many rigs now include a keyer as a standard option. You can plug the paddle into the radio, and you're on your way! CW operators tend to find paddle choice very personal, so definitely try one out before you buy. A hamfest often has one or more key–bug–paddle collectors, and you can try many styles. The ham behind the table is likely to be full of good information as well.

If you decide on an external keyer, you can choose kits or finished models. Programmable memories are very handy for storing commonly sent information, such as your call sign or a CQ message. Sometimes, I put my keyer in beacon mode to send a stored CQ message repeatedly to see whether anyone is listening on a dead band. (If everybody listens and nobody transmits, the band sounds dead but may be open to somewhere surprising.) Several computer programs send code from the keyboard. Browse to `www.ac6v.com/morseprograms.htm` for an extensive list of software.

A *voice keyer* is a device that can store short voice messages and play them back into your radio as though you were speaking. Some keyers are stand-alone units, and others use a sound card. Voice keyers are handy for contesting, DXing, calling CQ, and so on. Some models also store both CW and voice messages, such as the MJF Contest Keyer (www.mfjenterprises.com). Contest logging software such as N1MM (www.n1mm.com) and Writelog (www.writelog.com) can create a voice keyer by using the computer's sound card.

Antenna tuners

Antenna tuners don't really "tune" your antenna, but they allow your transmitter to operate at maximum efficiency no matter what impedance appears at the shack end of your feed line. Tuners are explained in the article "Do You Need an Antenna Tuner?" at www.arrl.org/transmatch-antenna-tuner.

Although your new radio may be equipped with an antenna tuner, in some situations you may need an external unit. Internal tuners have a somewhat limited range that fits many antennas. Antennas being used far from their design or optimized frequency often present an impedance that the rig's tuner can't handle. If balanced feed lines are used, you may need a tuner that can handle the change from coaxial cable to open-wire feed lines.

Tuners are available in sizes from tiny, QRP-size units to humongous, full-power boxes larger than many radios. Table 12-4 lists a few of the manufacturers offering an assortment of tuners. If you decide to purchase a tuner, choose one that's rated comfortably in excess of the maximum power you expect to use. I highly recommend getting one with the option to use balanced feed lines. The ability to switch between different feed lines and an SWR meter (which measures reflected RF power) is a nice-to-have feature.

Table 12-4 Antenna Tuner Manufacturers

Manufacturer	*Website*	*Balanced Feed Line*	*High-Power (>300 Watts)*	*Automatic Tuning*
MFJ Enterprises	www.mfjenterprises.com	Yes	Yes	Yes
Ameritron	www.ameritron.com	Yes	Yes	No
Vectronics	www.vectronics.com	Yes	Yes	No
LDG Electronics	www.ldgelectronics.com	External balun adapter	Yes	Yes
SGC	www.sgcworld.com	Yes	Yes	Yes

Along with the tuner, you need a *dummy load,* a large resistor that can dissipate the full power of your transmitter. The MFJ 260C can dissipate 300 watts, which is adequate for HF transceivers. High-power loads, such as the MFJ 250, immerse the resistor in a cooling oil. The dummy load keeps your transmitted signals from causing interference during tuneup. HF dummy loads may not be suitable for use at VHF or UHF, so check the frequency coverage specification before you buy.

Choosing a Computer for the Shack

A computer can be involved in almost every activity. Ham radio has embraced computers more intimately than most hobbies. Originally used as a replacement for the paper logbook, the computer in ham radio has evolved nearly to the point of becoming a second op, controlling radios, sending and receiving CW, and linking your shack to thousands of others through the Internet.

PC or Mac or . . . ?

Most ham-shack computers are Windows-based machines. The vast majority of software available for ham applications runs on the Windows operating systems.

Linux has an increasing number of adherents, particularly among digital-mode enthusiasts. Here are a couple of websites that focus on Linux software:

- **Hamsoft:** `http://radio.linux.org.au`
- **AC6V.com:** `www.ac6v.com/software.htm#LIN`

The Macintosh computing community is making inroads in ham radio software, and programs are available for all of the common ham radio uses available. The Ham-Mac mailing list (`www.mailman.qth.net/mailman/listinfo/ham-mac`) is full of information for Mac fans. A useful website devoted to bringing together Macintosh computers and ham radio is `http://machamradio.com`.

Regardless of what platform and operating system you prefer, software tools and programs are available to help you enjoy any type of operating you like. Some software is supplied by commercial businesses, and the amateur community has developed an amazing amount of shareware and freeware. Hams freely contribute their expertise in any number of ways, and developing software is a very popular activity.

Digital modes

Operation on most of the digital modes is rapidly converging on the sound card as the standard device to send and receive data. With a simple data and radio control interface, your computer and radio form a powerful data terminal. MFJ Enterprises (www.mfjenterprises.com) and West Mountain Radio (www.westmountainradio.com) both manufacture popular data interfaces.

If you choose to use an external multiple-mode controller for the digital modes, such as the Timewave PK-232 or DSP-599zx (www.timewave.com), Kantronics KAM (www.kantronics.com), or MFJ-1278B, you need only a terminal program such as Hyperterm, which is built into Windows.

Radio control

Radios have an RS-232 or USB control interface through which you can monitor and control nearly every radio function. (Icom uses a proprietary interface called CI-V that requires a converter.) Because of that flexibility, some control programs put the front panel on a computer. Some radio manufacturers have a radio control package that you can purchase or download. Third-party programs such as Ham Radio Deluxe (www.hrdsoftwarellc.com) integrate radio control with logging software.

Remote control

More and more hams are setting up stations and operating them by remote control. The most common way is to use the Internet with a PC at each location running a station control program. The remote station is typically configured something like Figure 12-7. (The additional DTMF controller connected to the phone line is needed in case the computer loses control of the radio and you need to shut everything down and start over.)

Why go to the trouble? Many subdivisions have restrictions against outside antennas, which means that you have to use indoor antennas in an attic, perhaps, or try to sneak in a thin wire around the yard. Either way, it's a compromise. Apartment and condo dwellers have similar problems. In addition, you can cause interference for your neighbors, and their electronic devices emit all kinds of noise, too. When you look at it that way, setting up a station out in the country sounds pretty good.

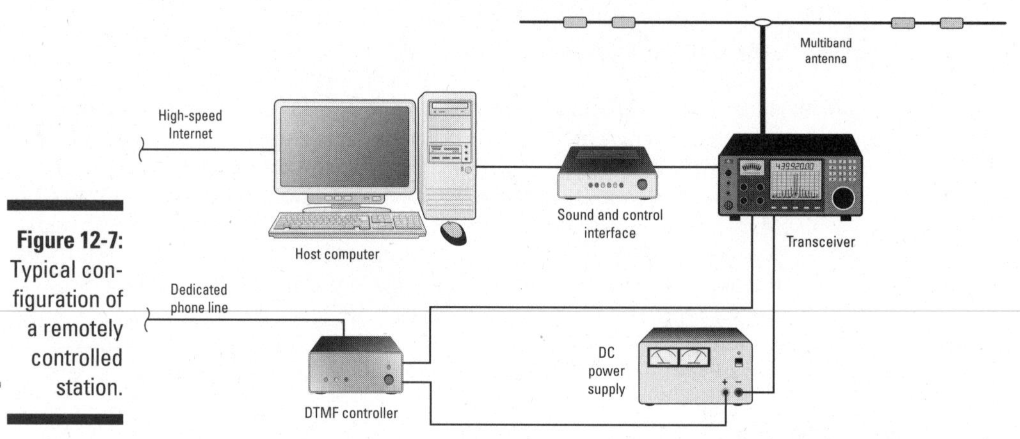

Figure 12-7: Typical configuration of a remotely controlled station.

Courtesy American Radio Relay League

Also, if you set up your station at home but operate it away from home, the controlling software can run on a laptop so you can fire up the rig from a hotel room or coffee shop. There's no reason why you have to lug around a laptop, either. A smartphone or tablet computer can be quite enough. The remote-control software developed by Pignology (`www.pignology.net`) runs on an iPhone and even includes logging software so that everything fits in the palm of your hand.

Regardless of how you do it, there are some rules about remote control operation, such as being able to shut down the station if necessary, being licensed in the location of the station, and identifying your station properly. Otherwise, it's pretty much like operating a regular station but with a *really* long microphone cord.

I expect remote operating to become widespread in the next few years. The technology to make it work is available, radios themselves are easier to control over the Internet, and commercial products are appearing that provide plug-and-play operation. You'll have the option of building a traditional home station and be able to operate it from wherever you are. As I say at the beginning of the chapter, this is an exciting time!

A good overview of the state of remote control at the time this edition was written (early 2013) is available in *Remote Operating for Amateur Radio,* by Steve Ford (WB8IMY). It's available from the ARRL and other ham radio booksellers. The *ARRL Handbook* also includes information on remote stations, including site evaluation and alternative power.

Hardware considerations

Beyond computation-intensive applications, such as antenna modeling or high-performance data modems, you don't need to own the latest and greatest speed-demon computer. If you're thinking about upgrading a home computer, a computer that's a couple of years old does just fine in the ham shack. Furthermore, the flood of cheap surplus computers available for a song means that you can dedicate a computer to its own specific task, such as running your logging software or monitoring an APRS website, so as not to tie up your main computer.

If you decide to purchase a new computer for the shack, be aware that the standard interface in ham radio for data and control remains the RS-232 serial COM port, which is being phased out on new computers in favor of USB 3.0. (RS-232 ports are now referred to as *legacy* ports.) Integrating a USB-only PC into the ham shack means that you either have to purchase a serial port expansion card or use USB-to-RS-232 converters. The serial-port expansion cards likely have fewer compatibility and driver issues, but the USB converters are easier to install. More and more radios and accessories are converting to USB interfaces, however.

Buying New or Used Equipment

New equipment is always safest for a neophyte, and it has that great new radio smell, too! If the equipment doesn't work, you have a service warranty, or the customer service representatives can help you out. Sales personnel can help you with information about how to set up a radio or any accessories, and may even have technical bulletins or application guides for popular activities. To find out where to buy new gear, get a copy of *QST, CQ,* or *WorldRadio.* The major dealers run ads every month, and some ads are virtual catalogs. You can find an online directory of ham radio stores, distributors, and manufacturers at `www.ac6v.com/hamdealers.htm`.

Although buying new equipment is safe, used gear is often an excellent bargain, and hams do love a bargain. You can find nearly any imaginable piece of gear with a little searching on online swap sites, including eBay. (Look for these sites through the portals and reflectors listed in Chapter 3.) I like to buy and sell through the ham radio websites. I like the Classified pages on the eHam.net portal (`www.eham.net/classifieds`), and QRZ.com also has for-sale forums. Enter *ham trade* or *ham swap* in an Internet search engine. As with shopping at hamfests, get help from an experienced friend before buying.

Internet and toll-free telephone numbers make shopping for rock-bottom prices easy, but you have to pay for shipping, which could add tens of dollars to your purchase. You may also have to pay for shipping to return equipment for service. Determine what happens if your new accessory needs repair before you buy.

A local electronics parts store is a valuable resource for you as well. If you are lucky enough to have a local ham radio dealer, even better! RadioShack probably doesn't carry replacement parts for your WhizBang2000, and it probably doesn't have a new battery pack for that 10-Band PocketMaster. When you're stuck in the middle of an antenna or construction project, you don't want to have to stop and wait for a courier to deliver materials. My advice: Buy some things online and some locally, balancing the need to save money against the need to support your local stores.

Upgrading Your Station

Soon enough, usually about five minutes after your first QSO, you start thinking about upgrading your station. Keep in mind the following tips when the urge to upgrade overcomes you. Remember the adage "You can't work 'em if you can't hear 'em!"

- The least expensive way to improve your transmit and receive capabilities is to use better antennas. Dollar for dollar, you get the most improvement from an antenna upgrade. Raise antennas before making them larger.
- Consider adding power only after you improve your antennas and eliminate local noise sources. Improve your hearing before extending the range at which people can hear you. An amplifier doesn't help you hear better at all.
- Buying additional receiving filters for an older radio is a whole lot cheaper than buying a new radio.
- The easiest piece of equipment to upgrade in the station is the multiple-mode processor between your ears. Before deciding that you need a new radio, be sure you know how to operate your old one to the best of your abilities. Improving your know-how is the cheapest and most effective improvement you can make.

By taking the improvement process one step at a time and by making sure that you improve your own capabilities and understanding, you can achieve your operating goals quicker and get much more enjoyment out of every ham radio dollar.

Chapter 13

Organizing Your Shack

In This Chapter

- Devising your ham shack layout
- Staying safe with RF and electrical currents
- Grounding your equipment

A well-organized shack provides many benefits for occasional and serious ham enthusiasts alike. You spend many hours in the shack, so why not make the effort to make your experience as enjoyable as possible? This chapter explains how to take care of the two most important inhabitants of the shack: the gear and you. The order of priority is up to you.

Designing Your Ham Shack

One thing you can count on is that your first station layout will prove to be unsatisfactory. It's guaranteed! Don't bolt everything down right away. Plan to change the layout several times as you change your operating style and preferences.

You'll spend a lot of time in your shack, no matter where it's located, so making it comfortable and efficient is important. By thinking ahead, you can avoid some common pitfalls and save money, too.

Keeping a shack notebook

Before you unpack a single box or put up one shelf in your shack, you should start a shack notebook to record how you put your station together and to help you keep your station operating. The notebook can be a simple spiral-bound notebook or a three-ring binder. A bound book has the advantage that its pages can't get blown out of the binder due to an unexpected gust of wind.

Sketch your designs before you begin building, make lists of equipment and accessories, and record the details as you go along. If you hook up two pieces of gear with multiconductor cable, write down the color of each wire and what it's attached to. You won't remember these details later, and the written list will save you tons of time and frustration.

After you have the station working, keep track of the gear you add and how you connect it. Record how your antennas work at different frequencies. Write down the wiring diagram for the little gadgets and adapters. Print and save any crucial software configuration information. Don't rely on memory! Taking a few minutes to record information saves tenfold the time later.

Building in ergonomics

Spending hours in front of a radio or workbench is common, so you need to have the same concerns about ergonomics in your radio shack that you do at work. You want to avoid awkward positions, too-low or too-high furniture, and harsh lighting, to name just a few. By thinking about these things in advance, you can prevent any number of personal irritations.

The focal point

Remember your main goals for the station. Whatever you plan to do, you'll probably use one piece of equipment more than half the time. That equipment ought to be the focal point of your station. The focal point can be the radio, a computer keyboard and monitor, or even a microphone or Morse code paddle. Paying attention to how you use that specific item pays dividends in operating comfort.

The computer

You may be building a radio station, but in many cases, you'll use your computer more than the radio. Certainly, monitors are the largest pieces of equipment at your operating position. Follow the guidelines for comfortable computer use. Position the desktop at the right height for extended periods of typing, or a keyboard tray might work. Buy a high-quality monitor, and place it at a height and distance for relaxed viewing. Figure 13-1 shows a few ways of integrating a monitor with radio gear.

Monitors mounted too far above the desk give you a sore neck. If you place the monitor too far away, your eyes hurt; if you place it too far left or right, your back hurts. Now is the time to apply computer ergonomics and be sure that you don't build in aches and pains as the reward for the long hours you spend at the radio.

If you wear glasses, get a special pair that focuses at the distance to the monitor. Your eyes will thank you!

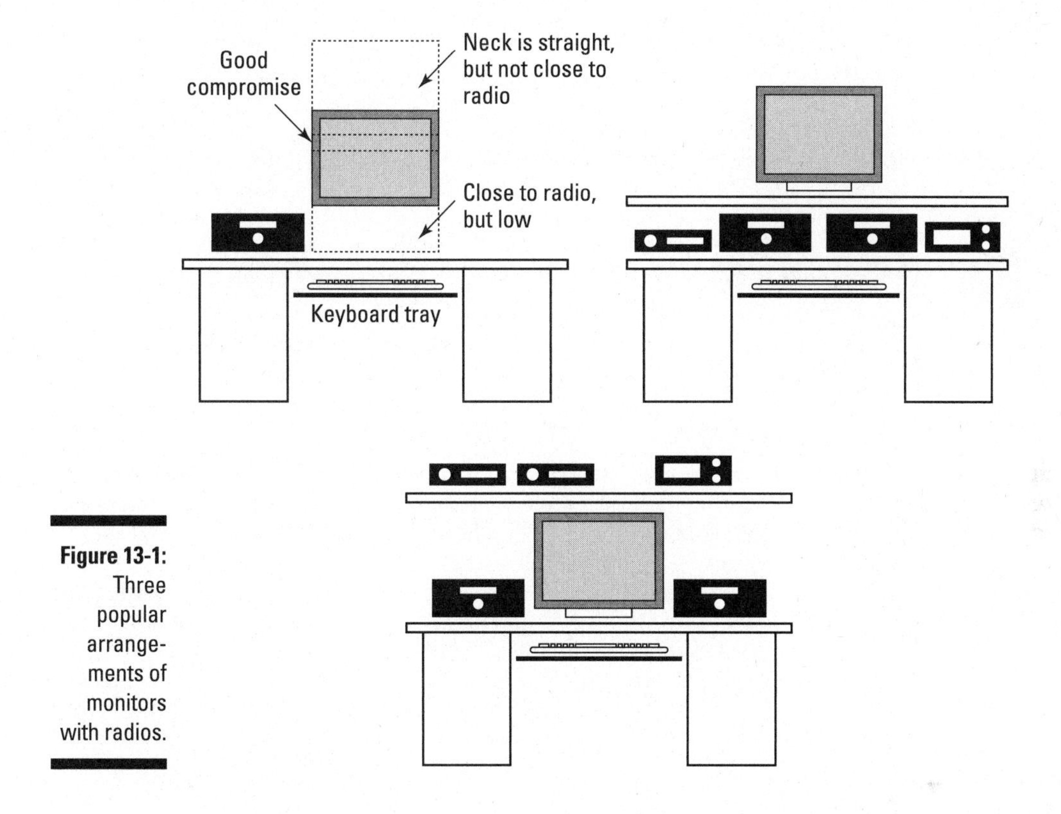

Figure 13-1: Three popular arrangements of monitors with radios.

The radio

Radios (and operators) come in all shapes and sizes, which makes giving hard-and-fast rules difficult. HF operators tend to do a lot of tuning, so placement of the radio is very important; VHF-FM operators do less tuning, so the radio doesn't have to be as close to the operator. Placing your most-used radio on one side of the keyboard or monitor probably is the most comfortable arrangement. If you're right-handed, have your main radio on your left and your mouse or trackball on your right. If you're left-handed, do the opposite.

Adjust the front of the radio to a comfortable viewing angle. You should be able to see all the controls and displays without having to move your head up or down.

The operating chair

A key piece of support equipment, so to speak, is the operator's chair. I'm always astounded to visit state-of-the-art radio stations and find cheap, wobbly, garage-sale office chairs at the operating positions. Even though the operator may have spent thousands of dollars on electronics, he or she doesn't get the most out of the radios because of the chair.

A good shack chair is a roll-around office-style chair with good lower-back support and plenty of padding in the seat. An adjustable model is best, preferably one that you can adjust with levers while sitting down. You may find that chairs with arms make sitting close to the operating desk difficult without leaning on your arms or stressing your lower back. If possible, remove the arms if you like sitting close to the operating table or desk.

You won't regret spending a little extra on the chair. Your body is in contact with your chair longer than with any other piece of equipment.

You can find good bargains on used chairs at used-office-furniture stores, listed in the Yellow Pages.

The desk and shelves

The top surface of your operating desk is the second-most-contacted piece of equipment. As is the case with chairs, many choices are available for desks suitable for a ham shack. Consider height and depth when looking at desks.

Before choosing a desk, you need to decide whether the radio will sit on the desk or on a shelf above it. Figure 13-2 illustrates the basic concerns. Do you like having your keyboard on the desk? You need to be comfortable sitting at your desk with your forearms resting comfortably on it. If you tune a radio a lot, such as for most HF stations, avoid arrangements that cause your arm to rest on the elbow or on the desk's edge for prolonged periods. Nothing is more painful! Make sure you have enough room for wrist support if using the keyboard is your main activity.

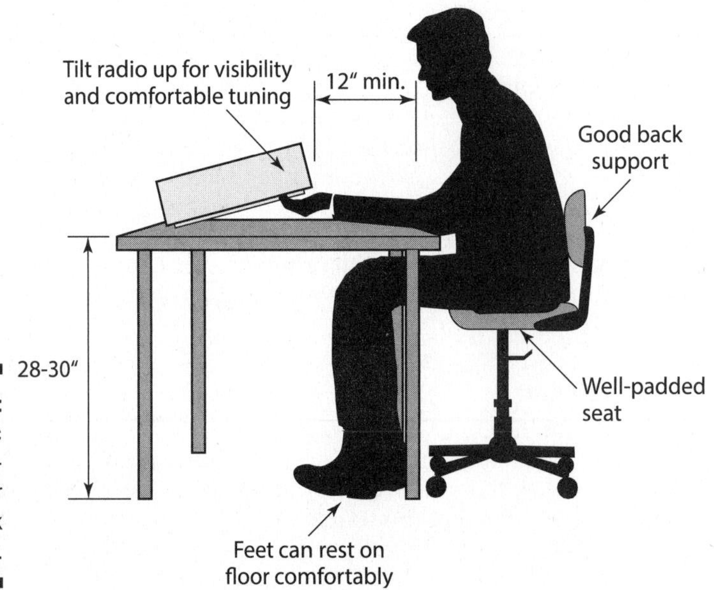

Figure 13-2: The basic considerations for shack desk and chair.

The most common height for desks is 28 to 30 inches from the floor. A depth of 30 inches is about the minimum if a typical transceiver that requires frequent tuning or adjustment is sitting directly on it. You need at least 12 inches between the front edge of the desk and the tuning knob. With your hand on the radio control you use most frequently, your entire forearm needs to be on the desktop. If the radio is sitting on a shelf above the desk, be sure that it's close enough to you that tuning is comfortable and doesn't require a long reach. Be sure you can see the controls clearly.

For small spaces, a computer workstation may be a good solution. You'll probably have to add some shelves, but the main structure has all the right pieces and may be adjustable to boot.

Viewing some examples

Because every station location and use is going to be different, the most helpful thing I can do is provide examples. Then you can decide what works for you. The goal of this section is to get you thinking about what works for you, not to suggest that you duplicate these stations exactly.

Ham shacks

Paul Beringer (NG7Z) faced the challenge of setting up a station in a condominium for low-power HF contesting and DXing. His solution, shown in Figure 13-3, was to use a home-assembled computer workstation with a fold-down front desk.

Beringer's workstation has these features:

- The radio and accessories are stacked where he can easily see and operate them.
- The fold-down surface provides enough desk surface for wrist support.
- A slide-away drawer holds the computer keyboard and mouse below the desk at a comfortable height.
- The monitor is close enough to be easy on the eyes and at a comfortable viewing angle even to the side of the radio.
- Lighting is from behind the operator to prevent glare.

Many hams like soft light in the shack because it's not distracting and makes reading the indicators and displays on the radios and accessories easy.

Figure 13-3: Paul Beringer (NG7Z) put a lot of contacts in his log from this compact layout.

The club station at the Missouri University of Science and Technology, WØEEE, has a larger area and more equipment. Club member Sterling Coffey (NØSSC) can operate on HF or VHF/UHF while another member uses the military-style equipment or chat on the local repeater (see Figure 13-4).

This setup has the following features:

- The computer monitor is right in the middle at a height that minimizes head and eye movement between the radios and software.
- Shelving holds larger gear, displays, and accessories above the rigs that are at tabletop level for easy tuning.
- There's lots of tabletop surface for the keyboard and mouse, making it easy and comfortable to operate without forearm or wrist strain.
- Everything that requires frequent adjustment is within easy reach. No side-to-side movement is required.

Figure 13-4: This multiple-operating-position station organizes gear by function.

Put paper labels on the front-panel tuning controls of an amplifier and antenna tuners to make changing frequencies easy without extended tuneups. This method minimizes stress on the amplifier and interference with other stations.

You needn't have gear stacked to the ceiling to have an effective station, as you see in the clean, simple layout of Don Steele (NTØF). Steele's specialty is operating in digital mode contests, usually on the RTTY (radioteletype) mode, which means a lot of monitor-watching. Figure 13-5 shows that the LCD computer monitor is the focus of the operating position and that the transceivers and all accessories are within easy reach.

Mobile and portable stations

Getting a radio installed in a car has never been easier, although the antenna installation can be challenging. In the past few years, manufacturers have taken the all-band radio design to new heights. One example is the Icom IC-7000. This radio measures only 6.6" × 2.3" × 7.9" and weighs only 5.1 pounds, but it's capable of developing 100 watts output on all amateur bands between

1.8 and 50 MHz, 50 watts on 144 MHz, and 35 watts on 432 MHz. This radio and similar offerings from other manufacturers have changed mobile and portable operation forever.

The compact size of modern radios allows for sleek installations in vehicles. You can even place radios with detachable front panels in the trunk or under a seat. As you can see in Figure 13-6, this arrangement provides lots of functionality in a small space. This mobile station can operate on all the HF bands and several VHF/UHF bands with all controls right at the driver's fingertips.

Connect the radio headphone output to your vehicle audio system's AUX input with a short stereo audio cable. You'll find high-fidelity mobile hamming to be easy on the ears.

Figure 13-5: This well-laid-out station reduces operator fatigue and looks good too.

Courtesy American Radio Relay League

Figure 13-6: Clever use of flexible mounts and cupholders makes for a compact mobile station.

Courtesy American Radio Relay League

If you plan to operate from your RV, take a few pointers from the clean layout shown in Figure 13-7, designed by Pete Wilson (K4CAV). When Wilson stops at a campground, he sets up a station right at the driver's position. The HF transceiver is not only mounted right on the dashboard, but also sitting on top of a mobile amplifier for an extra-powerful signal. When the RV is in motion, all this gear is safely stowed.

Figure 13-7: Pete Wilson (K4CAV) operates right from the driver's seat when his RV is parked.

Poorly secured radio gear can be *lethal* in an accident. Please don't drive without making sure your radios are securely mounted. Take care to keep cables and microphones away from airbag deployment areas as well. Anything in motion inside the vehicle is a hazard to you and your passengers.

Maybe you'd like something less tied to an automobile or RV. Well, how about a bicycle mobile station? The bicycle mobile station shown in Figure 13-8 belongs to Ben Schupack (NW7DX), who constructed it during his senior year of high school. Schupack uses a recumbent bike with an HF QRP rig mounted on the handlebars. A whip antenna and batteries are mounted behind the rider.

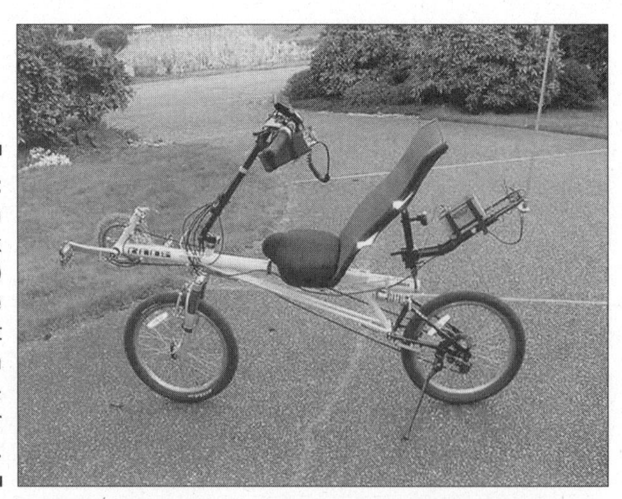

Figure 13-8: Ben Schupack (NW7DX) went for the recumbent look on his two-wheeler setup.

Building in RF and Electrical Safety

Whatever type of station you choose to assemble, you must keep basic safety principles in mind. Extensive literature is available for hams (see the sidebar "Sources of RF and electrical safety information," later in this chapter).

Don't think that you can ignore safety in the ham shack. Sooner or later, equipment gets damaged or someone gets hurt. Take a little time to review the safety fundamentals.

Basic safety

Safety isn't particularly complicated. Being safe consists mainly of consistently observing just a few simple rules. These tips can keep you out of more trouble than almost any others:

- **Know the fundamental wiring rules for AC power.** The National Electrical Code (NEC) contains the rules and tables that help you do a safe wiring job. The NEC, as well as numerous how-to and training references, is available in your local library or at home-improvement centers. If you're unsure of your skills, hire an electrician.
- **Deal with DC power carefully, especially in a car, to prevent short circuits and poor connections.** Either situation can cause expensive fires, and poor connections result in erratic operation of your radio. As with AC power, read the safety literature or hire a professional installer to do the job right.
- **Think of your own personal and family safety when constructing your station.** Don't leave any kind of electrical circuit exposed where someone can touch it accidentally. Use a *safety lockout* (a device that prevents a circuit breaker from being closed, energizing a circuit) on circuit breakers when you're working on wiring or equipment. Have fire extinguishers handy and in good working order. Show your family how to remove power from the ham shack safely.

Lightning

The power and destructive potential of lightning are awesome. Take the necessary steps to protect your station and home. These steps can be as simple as disconnecting your antenna feed lines when they're not in use. Alternatively, you may decide to use professional-level bonding and grounding. Whichever option you choose, do the job diligently and correctly.

RF exposure

The signals your transmitter generates can also be hazardous. The human body absorbs radio-wave frequency (RF) energy, turning it to heat. RF energy varies with frequency, being most hazardous in the VHF and UHF regions. A microwave oven operates at the high end of the UHF frequencies, for example.

Amateur signals are usually well below the threshold of any harmful effects but can be harmful when antennas focus the signal in such a way that you're exposed for a long period of time. High-power VHF and UHF amplifiers can definitely be hazardous if you don't handle them with caution.

A comprehensive set of RF safety guidelines is available. As you construct your own station, do a station evaluation to make sure you're not causing any hazards due to your transmissions.

First aid

As in any other hobby that involves the potential for injury, having some elementary skills in first aid is important. Have a first-aid kit in your home or shack, and be sure your family members know where it is and how to use it. Training in first aid and CPR is always a good idea for you and your family, regardless of your hobby.

Sources of RF and electrical safety information

Be responsible and check out these inexpensive safety references:

- The American Radio Relay League promotes safety for all manner of amateur radio activities. You can find excellent discussions of ham shack hazards and how to deal with them in the *ARRL Handbook* and the *ARRL Antenna Book*. Several articles are available in PDF format at www.arrl.org/safety.
- For safety issues relating to power circuits, the Electrical Safety Forum (www.electrical-contractor.net/ESF/Electrical_Safety_Forum.htm) is a good source of information.
- Brush up on lightning and grounding issues in a series of engineering notes on the Protection Group website (www.protectiongroup.com). Click the Knowledge Base link or search the site for *white papers* and *technical notes.*
- The ARRL publications *RF Exposure and You,* the *ARRL Handbook,* and the *ARRL Antenna Book* all discuss RF exposure safety issues.

Grounding Power and RF

Providing good grounding — for both AC and DC power as well as for the radio signals — is very important for a trouble-free ham shack. Although proper power grounding is fairly straightforward for AC and DC power, grounding for the radio signals in a ham station is a different problem altogether. You have to deal with both situations.

Grounding for AC and DC power

Most people hear the word *ground* and think "connected to the Earth." What the term really means, though, is the lowest common voltage for all equipment. The Earth is at zero voltage for power systems, for example. Grounding is really the process of making sure that different pieces of equipment have the same voltage reference.

In AC and DC power wiring, grounding provides safety by connecting exposed conductors, such as equipment cases, directly to the Earth or to a zero-voltage point, such as a building's metal structure. Guiding any current away from you in the event of a short circuit between the power source and the exposed conductors provides safety. When you keep all equipment at the same low voltage, no current flows between pieces of equipment if they touch each other either directly or if you touch both pieces simultaneously.

Power safety grounding uses a dedicated conductor — the so-called "third wire." In the home, three-wire AC outlets connect the ground pin to the home's ground at the master circuit-breaker box. Because the frequency of AC power is very low, the length of the ground wire doesn't matter. It must only be heavy enough to handle any possible fault currents.

In DC systems, because of the generally low voltages involved (less than 30 volts), power safety is less concerned with preventing shock than with minimizing excessive current and poor connections, which create a lot of heat and are significant fire hazards. You must pay careful attention to conductor size and keep connections tight and clean. As with AC power grounds, the length of the conductor usually isn't an issue, but its current-carrying capacity is.

Grounding and bonding at RF

The techniques that work for AC and DC power safety often don't work well for the high frequencies that hams use. For RF, a wire doesn't have to be very long before it starts acting like an antenna or transmission line. At 28 MHz, for example, an 8-foot piece of wire is about ¼ wavelength long. It can have high voltage on one end and very little voltage on the other.

Because constructing a wiring system that has one common low-voltage point at RF for all the equipment usually is impractical, station builders should avoid using the term *RF ground* in favor of the more general term *bonding,* which means keeping all equipment at the same RF voltage, not necessarily zero. You can bond equipment together at RF by connecting each piece of gear to a copper strap or pipe with a short piece of strap or wire, as shown in Figure 13-9. Ham gear usually has a ground terminal just for this purpose. Then connect the bonding strap or pipe to your AC safety ground rod with a heavy wire.

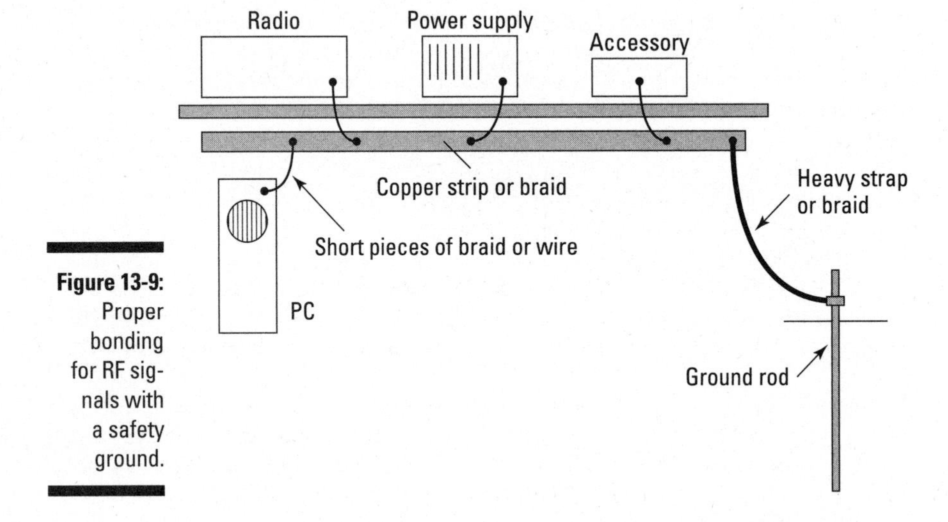

Figure 13-9: Proper bonding for RF signals with a safety ground.

Copper strap is sold at hardware and roofing stores as *flashing.* Avoid paying top dollar by finding a surplus-metals dealer and poking around. I was fortunate to find a heavy bar predrilled with evenly spaced holes that made a dandy bonding strip (see Figure 13-10). Use your imagination; all you need is material that's wide and easy to make good electrical contacts with.

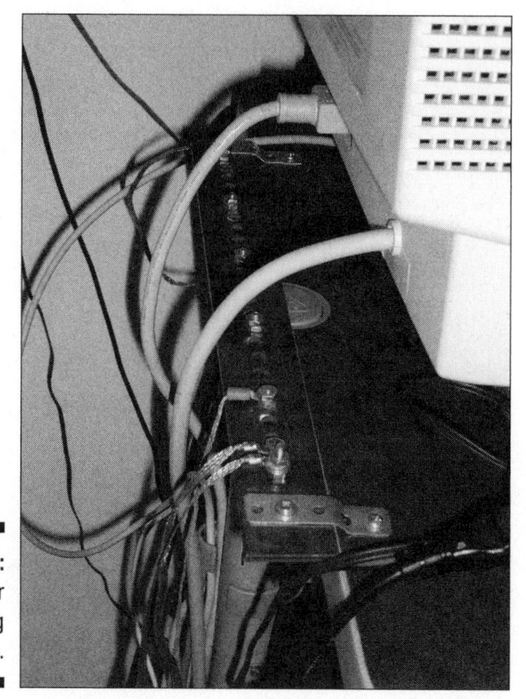

Figure 13-10: A copper bonding bus.

Chapter 14

Housekeeping (Logs and QSLs)

In This Chapter

- Maintaining a log
- Picking out your QSL card
- Mailing QSL cards

As you get ready for contacts with your new station, plan for the housekeeping chores that go with a new shack. For the ham, keeping a log of station operations and sending QSLs for contacts represent the paperwork that finishes the job. In this chapter, I show you what to put in your log and how to put together a QSL card.

Keeping a Log

Keeping a detailed log is no longer a requirement, but there are a lot of good reasons to keep track of what you do on the ham bands. The main log is a nice complement to the shack notebook (see Chapter 13), and you'll find it valuable in troubleshooting efforts and equipment evaluation.

Updating your little black radio book

Your station log can be a notebook or binder with handwritten entries for every contact. Figure 14-1 shows a typical format for paper logbooks.

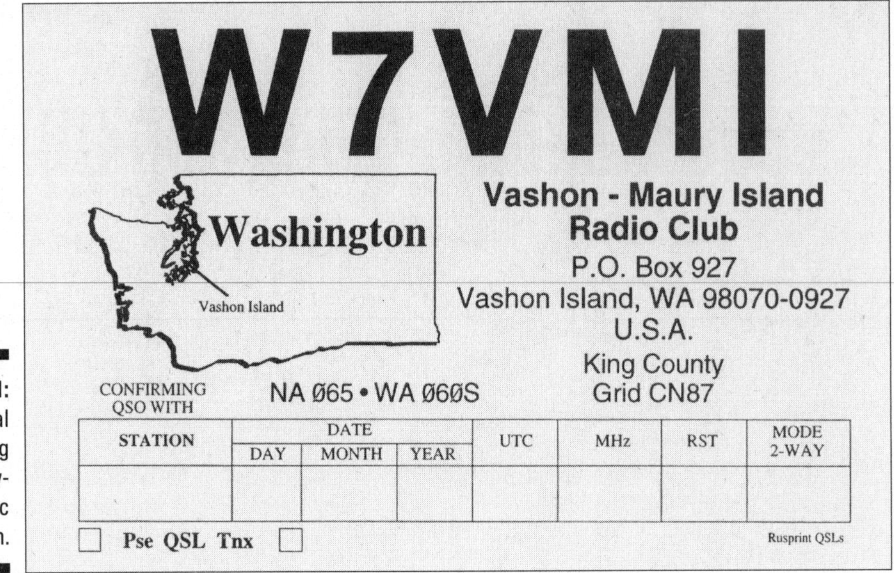

W7VMI

Washington

Vashon Island

Vashon - Maury Island
Radio Club
P.O. Box 927
Vashon Island, WA 98070-0927
U.S.A.
King County
Grid CN87

CONFIRMING QSO WITH NA Ø65 • WA Ø6ØS

STATION	DATE			UTC	MHz	RST	MODE 2-WAY
	DAY	MONTH	YEAR				

☐ Pse QSL Tnx ☐

Rusprint QSLs

Figure 14-1: A typical paper log sheet showing basic information.

Be sure to record the basics:

- **Time:** Hams keep time in UTC (or World Time) for everything but local contacts and radiogram identification (see Chapter 10).
- **Frequency:** Just recording the band in either MHz or wavelength is sufficient (20 meters or 14 MHz, for example).
- **Call sign:** Record each station you contact.

Those three pieces of information are enough to establish the who, when, and where of ham radio. Beyond the basics, you probably want to keep track of the mode you used, the signal report you gave and received, and any personal information about the other operator. Most people don't keep a log of casual local contacts made via repeaters, but you may want to log your participation in nets or training exercises. If so, it's okay to use local time.

Don't limit yourself to just exchanging the information recorded in your logbook or logging program. Another person is on the other end of the QSO with lots of interesting things to say.

Keeping your log on a computer

If you're an active ham, I highly recommend keeping your log on a computer. The logging program makes looking up previous contacts with a station or operator easy. You can also use your logging program like a web blog as a day-to-day radio diary to keep track of local weather, solar and ionospheric conditions, equipment performance, and behavior. (In fact, the word *blog* is a shortened version of *web log,* meaning "a log on the web.") The programs can also export data in standard file formats such as ADIF (Amateur Data Interchange Format). For mobile or portable operation, portable logging programs such as HamLog will run on a smartphone. InTheLog (`www.inthelog.com`) is a logging service "in the cloud" that you use with your web browser.

Selecting Your QSL Card

QSL cards, which are the size of standard postcards, feature an attractive graphic or photo with the station's call sign (see Figure 14-2). Information about a contact is written on one side. The QSL is mailed directly to the other station or through an intermediary. Exchanging QSL cards is one of my favorite activities, allowing you to claim credit for contacts to receive awards, but most hams do it just for fun and to build a lifetime collection.

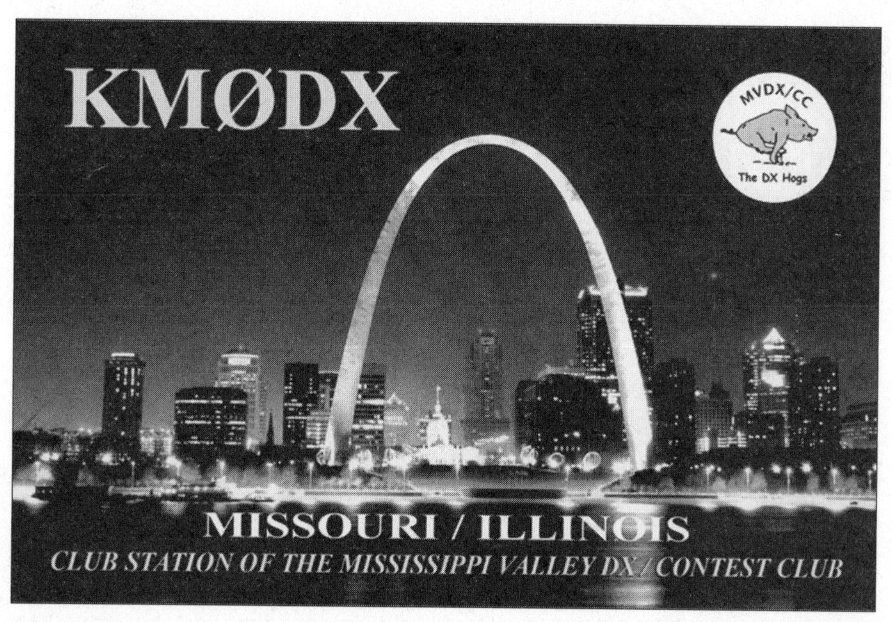

Figure 14-2: My club's QSL card.

Courtesy Mississippi Valley DX and Contest Club

You can find many varieties of QSLs, but three basic rules can help make the exchange as quick and error-free as possible:

- Have your call sign and QSO information on one side so that the receiver doesn't have to look for it.
- Print your call sign and all contact information in a clear, easy-to-read typeface or in capital letters.
- Beyond your mailing address, make sure that the QSL shows the physical location of the station, including county (for U.S. stations) and four-digit grid square.

You can find advertisements for QSL printers in ham magazines and on websites such as `www.ac6v.com/qslcards.htm`, or you can roll your own cards and have them printed at a local print or copy shop.

Follow these suggestions for surefire accuracy:

- **Double-check the dates and times of your contacts.** Date and time are frequent sources of error. Start by making sure that your own clock is set properly. Use UTC or World Time for every QSL except those for local contacts.
- **Use an unambiguous format for date.** Does 5/7/99 mean May 7 or July 5? The date is crystal-clear if you show the month with a Roman numeral, as in 7/V/99, or spell the month out, as in 5 Jul 99.
- **Use heavy black or blue ink** that won't fade over time. Never use pencil.

Sending and Receiving QSLs

After you fill out your QSLs, then what? Depending on where the cards are going, you have several options that weigh postage expense against turnaround time. By following the appropriate rules for each method, you can get a high return rate for your cards.

QSLing electronically

Although exchanging a paper card is fun (many hams send a card for the first contact with a station), more and more hams are confirming their contacts on two sites: eQSL and ARRL's Logbook of the World (LoTW).

eQSL (`www.eqsl.net`) was the first electronic QSL system and is extremely easy to use. Its site has a tutorial slideshow that explains just how eQSL works and how to use it.

The ARRL's LoTW (www.arrl.org/logbook-of-the-world) is more complicated to use. You're required to authenticate your identity and license, and all submitted contacts are digitally signed for complete trustworthiness. LOTW doesn't generate or handle physical QSL cards; it only provides verification of QSOs for award purposes, currently supporting ARRL awards and CQ's WPX award program, with more on the way.

DXpeditions often use an online QSLing system such as Club Log's OQSL system (http://secure.clublog.org). You can support the expeditioners with a donation and request your QSL at the same time. It's speedy, secure, and highly recommended.

Direct QSLing

If you want to send a paper card, the quickest (and most expensive) option is *direct,* meaning directly to other hams at their published addresses. You can find many ham addresses on the web portal QRZ.com. This method ensures that your card gets to recipients as fast as possible and usually results in the shortest turnaround time. Include the return postage and maybe even a self-addressed, stamped envelope. Direct QSLing costs more than electronic QSLing but makes it as easy as possible for you to get a return card on its way to the other ham — many times, with a colorful stamp.

Postal larceny can be a problem in poorer countries. An active station can make hundreds of contacts per week, attracting unwelcome attention when many envelopes start showing up with those funny number–letter call signs on them. Don't put any station call signs on the envelope if you have any question about the reliability of the postal service. Make your envelope as ordinary and as thin as possible.

Sending via managers

To avoid poor postal systems and cut postage expenses, many DX stations and nearly all DXpeditions use a QSL manager, which offers reliable, secure postal service and a nearly 100 percent return rate. QSLing via a manager is just like direct QSLing. If you don't include return postage and an envelope to a manager for a DX station, you'll likely get your card back via the QSL bureau (see the next section), which takes a few months at minimum. You can locate managers on websites such at QRZ.com's QSL Corner, which is free to members (www.qrz.com/page/qsl-corner.html), or the subscription GOlist (www.golist.net). The DX newsletters listed in Appendix B are also good sources of information.

If you send your QSL overseas, be sure to do the following:

- Use the correct global airmail letter rate from the U.S. Postal Service website (http://postcalc.usps.com).
- Ensure airmail service by using an Air Mail sticker (free at the post office), an airmail envelope, or an Air Mail/Par Avion stamp on the envelope.
- Include return postage from sources such as William Plum DX Supplies (e-mail plumdx@msn.com) or the K3FN Air Mail Postage Service (www.airmailpostage.com).

You may be asked to "send one (or two) greenstamps" for return postage. A *greenstamp* is a $1 bill. Take care that currency isn't visible through the envelope.

Bureaus and QSL services

All that postage can mount up pretty quickly. A much cheaper (and much slower) option exists: the QSL bureau system. You should use this method when the DX station says "QSL via the bureau" or, on CW, "QSL VIA BURO." The QSL bureau system operates as a sort of ham radio post office, allowing hams to exchange QSLs at a fraction of the cost of direct mail.

If you are an ARRL member, you can bundle up all your DX QSLs (you still have to send domestic cards directly) and send them to the outgoing QSL bureau, where the QSLs are sorted and sent in bulk to incoming QSL bureaus around the world. (The outgoing bureau is one of the best deals in ham radio.) The cards are then sorted and distributed to individual stations. The recipients send their reply cards back in the other direction. Go to www.arrl.org/qsl-service for more details. To get your cards, you must keep postage and envelopes in stock at your incoming QSL bureau. (Anyone can use the incoming QSL bureaus.) Then, when you least expect it, a fat package of cards arrives in the mail. What fun!

An intermediate route is the WF5E QSL Service (www.qsl.net/wf5e), which forwards QSLs to foreign and U.S. managers for a fee, currently 50 cents per card. You send outbound cards directly to WF5E, and your return cards come through the bureau system. Compared with direct mail, the WF5E QSL Service is a good bargain.

Chapter 15

Hands-On Radio

In This Chapter

- Acquiring tools and components
- Maintaining your station
- Troubleshooting
- Repairing your equipment
- Building equipment yourself

Ham radio is a lot more fun if you know how your radio works. You don't have to be an electrical engineer or a whiz-bang programmer, but to keep things running smoothly and deal with the inevitable hiccups, you need a variety of simple skills. As you tackle problems, you'll find that you're having fewer of them, getting on the air more, and having better luck making contacts. Trying new modes or bands will also be much easier for you.

To help you get comfortable with the hands-on part of ham radio, this chapter provides some guidance on the three parts of keeping a ham radio station on the air: making sure your radio doesn't break often, figuring out what's wrong when it does break, and fixing the broken part.

Before delving into the insides of your equipment, please take a minute to visit Chapter 13. Ham radio is a hobby, but electricity doesn't know that. I'd like to keep all my readers for a long, long time, so follow one of ham radio's oldest rules: Safety first!

The *ARRL Handbook* chapters on construction techniques and on troubleshooting are great references. The chapter on component data and references is full of useful tables and other data you'll need to refer to at the workbench.

Acquiring Tools and Components

To take care of your radio station, you need some basic tools. The job doesn't take a chest of exotic tools and racks of parts; in fact, you probably have most of the tools already. How many you need is really a question of how deeply you plan on delving into the electronics of the hobby. You have the opportunity to do two levels of work: maintenance and repair or building.

Maintenance tools

Maintenance involves taking care of all your equipment, as well as fabricating any necessary cables or fixtures to put it together. Figure 15-1 shows a good set of maintenance tools.

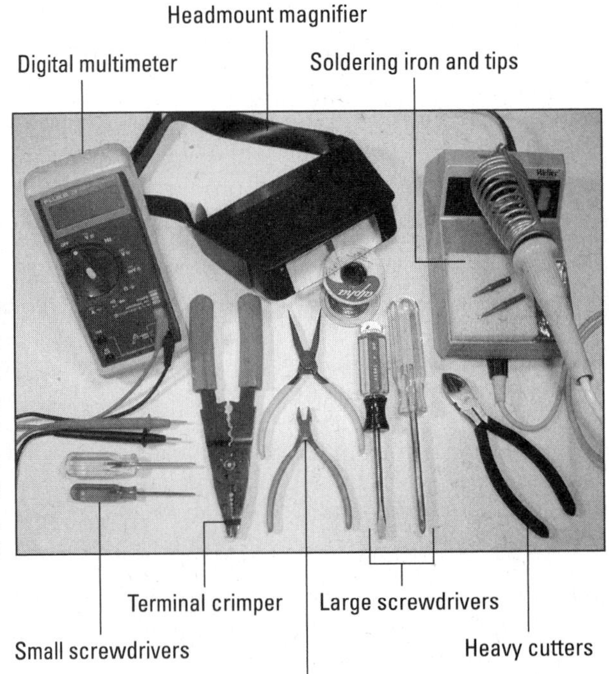

Figure 15-1: A set of tools needed for routine ham shack maintenance.

Having these tools on hand allows you to perform almost any electronics maintenance task:

- **Wire cutters:** Use a heavy-duty pair to handle big wires and cables, and a very sharp pair of diagonal cutters, or *dikes,* with pointed ends to handle the small jobs.
- **Soldering iron and gun:** You need a small soldering station with adjustable temperature and interchangeable tips. Delicate connectors and printed-circuit boards need a low-temperature, fine-point tip. Heavier wiring jobs take more heat and a bigger tip. A soldering gun should have at least 100 watts of power for antenna and cable soldering. Don't try to use a soldering gun on small jobs.
- **Terminal crimpers:** Regular pliers on crimp terminals don't do the job; the connection pulls out or works loose, and you spend hours chasing down the loose connections. Find out how to install crimp terminals and do the job right the first time with the right crimper for the terminal.
- **Head-mounted magnifier:** Electronic components are getting smaller by the hour, so do your eyes a favor. Magnifiers are often available at craft stores. You can also find clamp-mounted, swing-arm magnifier/light combinations.
- **Digital multimeter (DMM):** Even inexpensive models include diode and transistor checking, a continuity tester, and maybe a capacitance and inductance checker. Some models also include a frequency counter, which can come in handy.

Electronic components can be damaged by static electricity, such as when a spark jumps from your finger to a doorknob — a phenomenon called ESD (for *electrostatic discharge).* Inexpensive accessories for controlling ESD at your workplace and draining the static from your skin are available from RadioShack (`www.radioshack.com`) and other electronic-tool distributors.

You also need to have spare parts on hand. Start by having a spare for all your equipment's connectors. Look over each piece of gear and note what type of connector is required. When you're done, head down to the local electronics emporium and pick up one or two of each type. To make up coaxial cables, you need to have a few RF connectors of the common types. UHF, BNC, and N. SMA connectors, common on the newer handheld radios, take special tools to install. You'll purchase cables with SMA connectors already installed or adapters, as described next.

You often need adapters when you don't have just the right cable or a new accessory has a different type of connector. Table 15-1 shows the most common adapter types. You don't have to get them all at once, but this list is good to take to a hamfest or to use when you need an extra part to make up a minimum order.

Table 15-1 Common Shack Adapters

Adapter Use	Common Types
Audio	Mono to stereo phone plug (¼ inch and ⅛ inch), ¼ inch to ⅛ inch phone plug, right-angle phone plug, phone plug to RCA (phono) jack and vice versa, RCA double female for splices
Data	9-pin to 25-pin D-type, DIN-to-D cables, null modem cables and adapters, 9-pin and 25-pin double male/female (gender benders)
RF	Double-female (barrel) adapters for all four types of connectors, BNC plug to UHF jack (SO-239) and vice versa, N plug to UHF jack and vice versa, SMA to UHF adapter or jumper cable

A *plug* is the connector that goes on the end of a cable. A *jack* is the connector that's mounted on equipment. A *male connector* is one in which the signal contacts are exposed pins (disregard the outer shroud or shell). A *female connector* has recessed sockets that accept male connector pins.

Along with adapters and spare parts, you should have on hand some common consumable parts:

- **Fuses:** Have spares for all the fuse sizes and styles your equipment uses. Never replace a fuse with a higher-value fuse.
- **Electrical tape:** Use high-quality tape such as Scotch 33+ for important jobs, such as outdoor connector sealing, and get the cheap stuff for temporary or throwaway jobs.
- **Fasteners:** Purchase a parts-cabinet assortment with No. 4 through No. 10 screws, nuts, and lockwashers. Some equipment may require the smaller metric-size fasteners. You need ¼-inch and 5⁄16-inch hardware for antennas and masts.

Cleaning equipment is an important part of maintenance, and you need the following items:

- **Soft-bristle brushes:** Old paintbrushes (small ones) and toothbrushes are great cleaning tools. I also keep a round brush for getting inside tubes and holes.
- **Metal bristle brushes:** Light-duty steel and brass brushes clean up oxide and corrosion. Brass brushes don't scratch metal connectors but do damage plastic knobs or displays. Don't forget to clean corrosion or grease off a brush after the job.

- **Solvents and sprays:** I keep on hand a bottle or can of lighter fluid, isopropyl alcohol, contact cleaner, and compressed air. Lighter fluid cleans panels and cabinets gently and quickly, and also removes old adhesive and tape. Always test a solvent on a hidden part of a plastic piece before applying a larger quantity.

Repair and building tools

Figure 15-2 shows additional hand tools that you need when you begin doing your own repair work or building equipment. (The figure doesn't show larger tools such as drills and bench vises.)

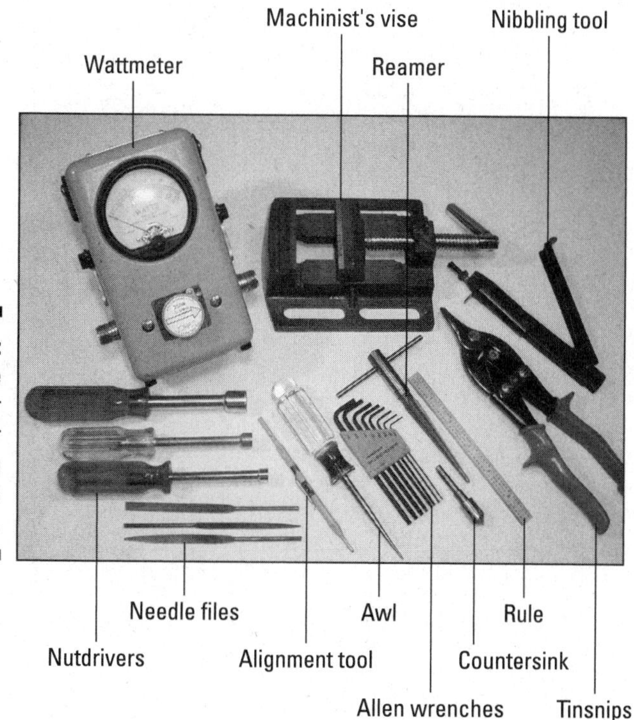

Figure 15-2: Use these tools for building or repairing electronic equipment.

Repairing and building go beyond maintenance in that you work with metal and plastic materials. You also need some additional specialty tools and instruments for making adjustments and measurements:

- **Wattmeter or SWR meter:** When troubleshooting a transmitter, you need an independent power-measurement device. Many inexpensive models work fairly well (stay away from those in the CB shops; they often aren't calibrated properly when not used on CB frequencies). If you do a lot of testing, the Bird Model 43 is the gold standard in ham radio. Different elements, or *slugs,* are used at different power levels and frequencies. You can find both used meters and elements.
- **RF and audio generators and oscilloscope:** Although you can do a lot with DC tests and a voltmeter, radio is mostly about AC signals. To work with AC signals, you need a way to generate and view them. If you're serious about getting started in electronics, go to `www.arrl.org/servicing-equipment` to read up on techniques and test equipment for working on radio equipment.
- **Nibbling tool and chassis punch:** Starting with a round hole, the nibbler is a hand-operated punch that bites out a small rectangle of sheet metal or plastic. Use a nibbler to make a large rectangular or irregular opening and then file the hole to shape. A stepped drill bit (for small holes) and a chassis punch (for big holes) makes a clean hole in up to ⅛-inch aluminum or 20-gauge steel. Chassis punches aren't cheap, but if you plan to build regularly, they can save you an enormous amount of time and greatly increase the quality of your work.
- **T-handled reamer and countersink:** The reamer allows you to enlarge a small hole to a precise fit. A countersink quickly smoothes a drilled hole's edges and removes burrs.

Drilling a hole in a panel or chassis that already has wiring or electronics mounted on or near it calls for special measures. You must prevent metal chips from falling into the equipment and keep the drill from penetrating too far. To control chips, put a few layers of masking tape on the side of the panel where you're drilling, with the outer layers kept loose to act as a safety net. To keep a drill bit from punching down into the wiring, use a small piece of hollow tubing that exposes just enough bit to penetrate all the way through.

Components for repairs and building

I find myself using the same components in the following list for most building and repair projects. Stock up on these items (assortments are available from component sellers), and you'll always have what you need:

- **Resistors:** Various values of 5 percent metal or carbon film, ¼- and ½-watt fixed-value resistors; 100, 500, 1k, 5k, 10k, and 100k ohm variable resistors and controls

- **Capacitors:** 0.001, 0.01, and 0.1 μF ceramic; 1, 10, and 100 μF tantalum or electrolytic; 1000 and 10000 μF electrolytic; miscellaneous values between 220 pF and 0.01 μF film or ceramic
- **Inductors:** 100 and 500 μH, and 1, 10, and 100 mH chokes
- **Semiconductors:** 1N4148, 1N4001, 1N4007, and full-wave bridge rectifiers; 2N2222, 2N3904, and 2N3906 switching transistors; 2N7000 and IRF510 FETs; red and green LEDs
- **ICs:** 7805, 7812, 78L05, and 78L12 voltage regulators; LM741, LM358, and LM324 op-amps; LM555 timer; LM386 audio amplifier

Although having a completely stocked shop is nice, you'll find that building up the kinds of components you need takes time. Rather than give you a huge shopping list, I'll give you some guidelines to follow:

- **When you buy or order components for a project, order extras.** The smallest components — such as resistors, capacitors, transistors, and diodes — are often cheaper if you buy in quantities of ten or more. After a few projects, you have a nice collection.
- **Hamfests are excellent sources of parts and component bargains.** Switches and other complex parts are particularly good deals. Parts drawers and cabinets often come with parts in them, and you can use both.
- **Broken appliances and entertainment devices around the home are worth stripping before throwing out.** Power cords and transformers, headphone and speaker jacks, switches, and lots and lots of other interesting hardware items end up in the dump. Also, seeing how these items are made is interesting.
- **Build up a hardware junk box by tossing in any loose screws, nuts, spacers, springs, and so on.** Use an old paint tray or a flat open tray to make it easy to root through the heap in search of a certain part. The junk box can be a real time and money saver.

Sporting-goods and crafts stores have frequent sales on inexpensive tackle boxes and multiple-drawer cases that are perfect for electronic components.

Maintaining Your Station

The best thing you can do for your station is to spend a little time doing regular maintenance. Maintenance works for cars, checkbooks, and relationships, so why not ham radio?

Be sure to keep a station notebook (see Chapter 13). Open the notebook whenever you add a piece of equipment, wire a gadget, note a problem, or fix a problem. Over time, the notebook helps you prevent or solve problems, but only if you keep it up to date.

You also need to set aside a little time on a regular basis to inspect, test, and check the individual components that make up the station. Along with the equipment, check the cables, power supplies, wires, ropes, masts, and everything else between the operator and the ionosphere. Check these items when you plan to be off the air so that you don't have to do a panic fix when you want to be on the air. Your equipment and antennas are of no use if they're not working.

You can make routine maintenance easy with a checklist. Start with the following list and customize it for your station:

- **Check all RF cables, connectors, switches, and grounds.** Make sure all connectors are tight because temperature cycles can work them loose. Rotate switches or cycle relays to keep contacts clean and turn up problems. Look for kinks in or damage to feed lines. Be sure that ground connections are snug.
- **Test transmitters and amplifiers for full power output on all bands.** Also, double-check your antennas and RF cabling. Use full power output to check all bands for RF feedback or pickup on microphones, keying lines, or control signals.
- **Check received noise level (too high or too low) on all bands.** The noise level is a good indication of whether feed lines are in good shape, preamps are working, or you have a new noise source to worry about.
- **Check SWR on all antennas.** Be especially vigilant for changes in the frequency of minimum SWR, which can indicate connection problems or water getting into the antenna or feed line connectors. Sudden changes in SWR (up or down) mean tuning or feed-line problems. SWR is discussed in Chapter 12.
- **Inspect all antennas and outside feed lines.** Use a pair of binoculars to check the antenna. Look for loose connections; unraveling tape, ties, or twists; damage to cable jackets; and that sort of thing.
- **Inspect ropes and guy wires.** Get into the habit of checking for tightness and wear whenever you walk by. A branch rubbing on a rope can eventually cause a break. Knots can come loose.
- **Inspect masts, towers, and antenna mounts.** The best time to find problems is in autumn, before the weather turns bad. Use a wrench to check tower and clamp bolts and nuts. Fight rust with cold galvanizing spray paint. In the spring, check again for weather damage.

- **Vacuum and clean the operating table and equipment; clear away loose papers and magazines.** Sneak those coffee cups back to the kitchen, and recycle the old soft-drink cans. Make sure that all fans and ventilation holes are clean and not blocked.

  I realize that you may not want to haul the vacuum cleaner into the radio shack, but it may be the most valuable piece of maintenance gear you have. Heat is the mortal enemy of electronic components and leads to more failures than any other cause. The dust and crud that settle on radio equipment restrict air flow and act as insulators, keeping equipment hot. High-voltage circuits, such as in an amplifier or computer monitor, attract dust like crazy. Vacuuming removes the dust, wire bits, paper scraps, and other junk before they cause expensive trouble.

As you complete your maintenance, note whether anything needs fixing or replacing and why, if you know. You'll probably get some ideas about improvements or additions to the station, so note those ideas too.

Over time, you'll notice that some things regularly need work. In my mobile station, the antenna mounts need cleaning, vibration loosens connectors, and cables can get pinched or stretched. I'm always on the lookout for these problems.

If you do routine maintenance three or four times a year, you can dramatically reduce the number of unpleasant surprises you receive.

Overall Troubleshooting

No matter how well you do maintenance, something eventually breaks or fails. Finding the problem quickly is the hallmark of a master, but you can become a good troubleshooter by remembering a few simple rules:

- **Try not to jump to conclusions.** Work through the problem in an orderly fashion. Write your thoughts down to help focus.
- **Start with the big picture.** Work your way down to equipment level.
- **Avoid making assumptions.** Check out everything possible for yourself.
- **Read the equipment owner's manual, and get a copy of the service manual.** The manufacturer knows the equipment best.
- **Consult your station notebook.** Look for recent changes or earlier instances of related behavior.
- **Write down any changes or adjustments you made while troubleshooting so you can reverse them later.** You may not remember everything.

Troubleshooting Your Station

Your station is a system of equipment and antennas. To operate properly, each piece of equipment expects certain signals and settings at each of its connectors and controls. You can trace many station problems to those signals and settings, often without using any test equipment more sophisticated than a voltmeter.

Most station problems fall into two categories: RF and operational. RF problems are things such as high SWR, no signals, and reports of poor signal quality. Operational problems include not turning on (or off) properly, not keying (or keying inappropriately), or no communications between pieces of equipment.

Start by assigning the problem to one of these categories. (You may be wrong, but you have to start somewhere.)

RF problems

Some RF problems occur when RF isn't going where it's supposed to go. These problems generally are caused by a bad or missing cable, connector, or switching device (a switch or relay) that needs to be replaced. Try fixing these problems with the following suggestions:

- Replace cables and adapters one at a time, if you have spares that you know work.
- Note which combinations of switching devices and antennas seem to work and which don't. See whether the problem is common to a set or piece of equipment or specific cables.
- Bypass or remove switches, relays, or filters. Leave yourself a note to put the device back in.
- Check through antenna feed lines. Take into account whether the antenna feed point has a DC connection across it, such as a tuning network or impedance-matching transformer. Gamma-matched Yagi beams show an open circuit, whereas beta-matched Yagis and quad loops have a few ohms of resistance across the feedpoint. (***Note:*** Recording the normal value of such resistances in the station notebook for comparison when troubleshooting is a good idea.)

Other problems you may come across include "RF hot" microphones and equipment enclosures, and interference to computers or accessories. (You haven't fully lived until you get a little RF burn on your lip from a metal microphone case!) Usually, you can fix these problems by bonding equipment together (see Chapter 13). Try these suggestions:

- Double-check to ensure that the equipment is connected to the station RF ground bus. The equipment may be connected, but double-checking never hurts.
- Check the shield connections on audio or control cables. These cables are often fragile and can break when flexed or yanked. (You never yank cables, do you?)
- Change the location of the bonding wire, or coil up an excessively long cable.
- Add ferrite RF suppression cores to the cables (see the "Ferrites as RFI suppressors" sidebar, later in this chapter).

On the higher HF bands (particularly 21, 24, and 28 MHz), cables and wires begin to look like antennas as their lengths exceed ⅛ wavelength. A 6-foot data cable, for example, is about 3⁄16 wavelength long on 28 MHz and can have a sizable RF voltage at the midpoint, even though both ends are connected to the station's RF bus. If you have RF pickup problems on just one band, try attaching a ¼-wavelength *counterpoise* wire to move the RF hot spot away from the equipment in question. A ¼-wavelength wire left unconnected at one end can look short-circuited at the other end. Attaching the counterpoise to the enclosure of the affected equipment may lower the RF voltage enough to reduce or eliminate the interference. Keep the wire insulated and away from people and equipment at the unconnected end.

Power problems

Power problems can be obvious (no power), spectacular (failure of the high-voltage power supply), or subtle (AC ripple, slightly low or high voltage, or poor connections). The key is to never take power for granted. Just because the power supply light is on doesn't mean the output is at the right voltage. I've wasted a lot of time due to not checking power, and now I always check the power supply voltages first. Try these solutions to fix your power problems:

- **Check to see whether the problem is caused by the equipment, not the power supply.** You can easily isolate obvious and spectacular failures, but don't swap in another supply until you're sure that the problem is, in fact, the power supply. Connecting a power supply to a shorted cable or input can quickly destroy the supply's output circuits. If a circuit breaker or fuse keeps opening, don't jumper it. Find out why it's opening.
- **Check for low output voltage.** Low voltage, especially when transmitting, can cause radios to exhibit all sorts of strange behavior. The microprocessor may not function correctly, leading to bizarre displays, loss of external control, and incorrect response to controls. Low voltage can also result in low power output or poor RF stability (chirpy, drifting, or raspy signals).

- **Check the supply with both AC and DC meter ranges.** Hum on your signal can mean a failing power supply or battery. A DC voltmeter check may be just fine, but power supply outputs need to show less than 100 mV of AC.
- **If you suspect a poor connection, measure voltage at the load (such as the radio) and work your way back to the supply.** Poor connections in a cable or connector cause the voltage to drop under load. They can be difficult to isolate because they're problematic only with high current, such as when you're transmitting. Voltage may be fine when you're just receiving. Excessive indicator-light dimming is a sure indicator of poor connections or a failing power supply.

Working on AC line-powered and 50-volt or higher supplies can be dangerous. Follow safety rules, and get help if you're unsure of your abilities.

If your USB device is powered from the USB host, be sure that the host can supply power at the required amount of current. Remember that portable USB hubs often don't supply power unless connected to an AC adapter. Similarly, a laptop may be configured not to supply USB power unless its battery charger is working.

Operational problems

Operational problems fall into three categories: power, data, and control. After you determine which type of problem you have, you often come very close to identifying the cause of the problem.

Data problems

Data problems are more and more common in modern radio shacks. Interfaces among computers, radios, and data controllers usually are made with RS-232 or USB connections. Internet-connected equipment uses Ethernet or Wi-Fi networks. If you installed new equipment and can't get it to play with your other equipment, four common culprits are to blame:

- **Baud rate:** An improper *baud rate* (or the data framing parameters of start bits, stop bits, and parity of the RS-232 link) renders links inoperative, even if the wiring is correct. Baud rate specifies how fast data is sent. The framing parameters specify the format for each byte of data. These parameters are usually set by a menu or software configuration.
- **Protocol errors:** Protocol errors generally result from a mismatch in equipment type or version. A program using a Kenwood radio-control protocol can't control a Yaesu or Ten-Tec radio, for example. Be sure

that all the equipment involved can use the same protocol or is specified for use with the exact models you have.

- **Improper wiring configuration:** Be sure that you used the right cables. A null modem RS-232 cable or a crossover network cable may be required.
- **Network problems:** These problems are in a class of their own, but the equipment generally has a configuration or setup procedure that you can perform or review to see whether you have these problems.

If equipment that was communicating properly suddenly fails, you may have a loose cable, or the configuration of the software on one end of the link may have changed. Double-check the communications settings, and inspect the connections carefully.

USB interfaces go through a process of establishing a connection when the cable is connected. On a computer, icons indicate that the equipment is recognized and working properly (or not). On stand-alone equipment, you may see indicator lights change or icons on a display. Check the user manual, and watch for these changes carefully.

Control problems

Control problems are caused by either the infamous pilot error (in other words, you) or actual control input errors.

Pilot error is the easiest, but most embarrassing, type to fix. With all the buttons and switches in the shack, I'm amazed that I don't have more problems. Follow these steps to fix your error:

1. **Check that all the operating controls are set properly.**

   Bumping or moving a control by accident is easy. Refer to the operator's manual for a list of settings for the various modes. Try doing a control-by-control setup, and don't forget the controls on the back panel or under an access panel.

   Speaking from hard personal experience, before you decide that a radio needs to go to the shop, check every control on the front panel, especially squelch (which can mute the audio), MOX (which turns the transmitter on all the time), and Receive Antenna (which makes the receiver sound dead if no receive antenna is attached). If you're really desperate, most radios have the capability to perform a *hard reset,* which restores all factory default settings but also wipes out the memory settings.

2. **Disconnect every cable from the radio one at a time, except for power and the antenna.**

Start with the cable that contains signals related to the problem. If the behavior changes for any of the cables, dig into the manual to find out what that cable does. Could any of the signals in that cable cause the problem? Check the cable with an ohmmeter, especially for intermittent shorts or connections, by wiggling the connector while watching the meter or listening to the receiver.

3. **If the equipment isn't responding to a control input, such as keying or PTT, you need to simulate the control signal.**

   Most control signals are switch or contact closures between a connector pin and ground or 12V. You can easily simulate a switch closure with . . . a switch! Replace the control cable with a spare connector, and use a clip lead (a wire with small alligator clips on each end) to jumper the pin to the proper voltage. You may want to solder a small switch to the connector with short wires if the pins are close together. Make the connection manually, and see whether the equipment responds properly. If so, something is wrong in the cable or device generating the signal. If not, the problem is in the equipment you're testing.

At this point, you'll probably have isolated the problem to a specific piece of equipment, and your electronics skills can take over. You have a decision to make. If you're experienced in electronics and have the necessary information about the equipment (schematic or service manual), by all means go ahead with your repairs. Otherwise, proceed with caution.

Troubleshooting Your Home and Neighborhood

If you have problems outside your shack, they usually consist of dreaded RF interference (RFI), as in "I can hear you on my telephone!" or "My garage door is going up and down!" Less known, but just as irritating, is the man-bites-dog situation, in which your station receives interference from some other electric or electronic device. Solving these problems can lead you through some real Sherlock Holmes-ian detective work.

Start by browsing the ARRL RFI Information page at `www.arrl.org/radio-frequency-interference-rfi`. For in-depth information, including diagrams and how-to instructions, read *The ARRL RFI Book,* which covers every common interference problem. Your club library may have a copy. Consult your club experts for assistance. Occasional interference problems are facts of life in this era, and you're not the only ham who experiences them. Draw on the experience and resources of other hams for help.

Dealing with interference to other equipment

Start by making your own home interference-free. Unless you're a low-power VHF/UHF operator, you likely own at least one appliance that reacts to your transmissions by buzzing, humming, clicking, or doing its best duck imitation when you're speaking. It's acting like a very unselective AM receiver, and your strong signal is being converted to audio, just like the old crystal radio sets did. It's not the ham radio's fault — the appliance is failing to reject your signal — but it's still annoying.

Your goal is to keep your signal out of the appliance so that it doesn't receive the signal. Sounds simple, doesn't it? Start by removing all accessory cords and wires to see whether the problem goes away. If it does, put the cords and wires back one at a time to see which one is acting as the antenna. Power cords and speaker leads are very good antennas and often conduct the RF into the appliance. Wind candidate cables onto a ferrite interference suppression core (see the sidebar "Ferrites as RFI suppressors," later in this chapter) close to the appliance to find out whether that cures the problem. You may have to add cores to more than one of the leads, although generally just one or two are sensitive. RadioShack sells AC power cord filters (part 15-1111B) that may help.

If the device is battery-powered and doesn't have any leads, you probably can't fix the problem, I'm sorry to say. You either have to replace the device or get along with the interference. The manufacturer's website may have some interference cures, or you may find some guidance from ham radio websites or club members. Try entering the model number of the appliance and *interference* or *RFI* in a search engine to see what turns up.

The stored messages in the archives of the RFI e-mail reflector at `www.contesting.com` are good sources of information. You don't have to be a member of the reflector to read them. Start by searching for messages about the type of device you're having trouble with. If you find information about it, narrow your search with additional terms.

The following common devices are often victims of interference:

- **Cordless telephones:** Older phones that use 47 MHz frequencies are often devastatingly sensitive to strong signals. Luckily, newer phones use 900 MHz and 2.4 GHz radio links and are much less sensitive to your RF. If you come up against one of the 47 MHz units, just replace it with a newer one.

- **Touch lamps:** These accursed devices respond to nearly any strong signal on any frequency. You can try ferrite cores on the power cord, but results are definitely mixed. Internal modifications are described on the ARRL RFI website. Replacing the lamp may be the easiest option.
- **TV, video, and audio equipment:** A common path for interference is via the speaker wires, but any of the many connections among pieces of equipment can be picking up RF. Make sure that all the equipment ground terminals are connected by short, stout wires. The ARRL's RFI website is probably your next stop.
- **Alarm systems:** The many feet of wire strung around the house to the various sensors and switches create a dandy antenna. Unfortunately, the system controller sometimes confuses the RF that these wires pick up for a sensor signal. System installers have factory-recommended interference suppression kits that take care of most problems.
- **Switching power supplies:** Miniature power supplies that mount directly on the AC outlet (called *wall warts)* use rapidly switching electronics to convert AC power to low-voltage DC. Although these devices are efficient, they can be real noisemakers. Sometimes, a ferrite core on the DC output helps, but the best solution is to replace the supply with a regular linear supply that uses a transformer. Battery chargers sometimes use switching electronics, too, and they're much harder to fix or replace. You can often get relief by turning them off or unplugging them when the batteries are fully charged.

By practicing on your own home electronics, you gain valuable experience in diagnosing and fixing interference problems. Also, if a neighbor has problems, you're prepared to deal with the issue. See the nearby sidebar, "Part 15 devices."

Part 15 devices

Unlicensed devices that use RF signals to operate or communicate are subject to the Federal Communications Commission's Part 15 rules. These rules apply to cordless phones, wireless headphones, garage-door openers, and other such devices. Devices that may radiate RF signals unintentionally, such as computers and videogame consoles, are also subject to Part 15 rules. The rules make a tradeoff: Device owners don't need a license to operate, say, a cordless phone, but they must not interfere with licensed stations (such as ham radio stations), and they have to accept interference from licensed stations. This agreement generally works pretty well except in the strong-transmitter/sensitive-receiver neighborhood of a ham radio station. See the extensive discussion of Part 15 rules at `www.arrl.org/radio-frequency-interference-rfi` for more details.

Dealing with interference to your equipment

Two noise sources are likely to cause interference: electric and electronic. *Electric noise* is caused primarily by arcing in power lines or equipment, such as motors, heaters, and electric fences. *Electronic noise* is caused by leaking RF signals from consumer appliances and computers operating nearby. Each type has a distinctive *signature,* or characteristic sound. The following list describes the signatures of common sources of electric noise:

- **Power line:** Steady or intermittent buzzing at 60 Hz or 120 Hz. The weather may affect interference.

  Power-line noise is caused by arcing or corona discharge. *Arcing* can occur around or even inside cracked or dirty insulators. It can also occur when two wires, such as neutral and ground wires, rub together. *Corona discharge* occurs at high-voltage points on sharp objects where the air molecules become ionized and electricity leaks into the atmosphere. The interference is a 120 Hz buzzing noise because the arc or discharge occurs at the peaks of the 60 Hz voltage, which occur twice per cycle.

  Do not attempt to fix problems with power lines or power poles. *Always* call your power company.

  You can assist the power company by locating the faulty equipment. You can track down the noise source with a battery-powered AM radio or VHF/UHF handheld radio with an AM mode (aircraft band works well). If you have a rotatable antenna at home, use it to pinpoint the direction of the noise. (The null off the side of a beam antenna is sharper than the peak of the pattern.) Walk or drive along the power lines in that direction to see whether you can find a location where the noise peaks. I've found several power poles with bad hardware by driving around with the car's AM radio tuned between stations. If you do find a suspect pole, write down any identifying numbers on the pole. Several numbers for the different companies that use the pole may be on it; write them all down. Contact your utility and ask to report interference. You can find a great deal more information about this process on the ARRL RFI web page (`www.arrl.org/radio-frequency-interference-rfi`).

- **Industrial equipment:** Sounds like power-line noise but with a more regular pattern, such as motors or heaters that operate on a cycle. Examples in the home include vacuum cleaners, furnace fans, and sewing machines.

- **Defective contacts:** Highly erratic buzzes and rasps, emitted by failing thermostats or switches carrying heavy loads. These problems are significant fire hazards in the home, and you need to fix them immediately.

- **Dimmers and speed controls:** Low-level noise like power lines that comes and goes as you use lights or motors.

- **Vehicle ignition noise:** Buzzing that varies with engine speed, which is caused by arcing in the ignition system.
- **Electric fences:** Regular pop-pop-pop noises at about 1-second intervals. A defective charger can cause these problems, but the noise is usually due to broken or missing insulators or arcing from the fence wires to weeds, brush, or ground.

Finding an in-home source of electric noise depends on whether the device is in your home or a neighbor's. Tracking down in-home sources can be as simple as recognizing the pattern when the noise is present and recognizing it as the pattern of use for an appliance. You can also turn off your home's circuit breakers one at a time to find the circuit powering the device. Then check each device on that circuit.

If the noise is coming from outside your home, you have to identify the direction and then start walking or driving with a portable receiver. Review the ARRL RFI website or reference texts for information about how to proceed when the interfering device is on someone else's property.

What about electronic noise? The following list describes the signatures of common sources of electronic noise:

- **Computers, videogame consoles, and networks:** These devices produce steady or warbling tones on a single frequency that are strongest on HF, but you can also hear them at VHF and UHF.
- **Cable and power-line modems:** You hear steady or warbling tones or hissing/rasping on the HF bands.
- **Cable TV leakage:** Cable TV signal leakage at VHF and UHF sounds like a buzzing (video signal) or audio FM signal with program content. Cable channel 12 covers the same frequencies as the 2 meter band, for example. Cable TV systems are converting to digital signals that just sound like hissing noise.
- **Plasma TVs:** Although a few models are RF-quiet, many generate noise across a wide spectrum of frequencies. The only solution seems to be to replace them with LCD or LED models, which don't have the noise problem.

Each type of electronic interference calls for its own set of techniques for finding the source and stopping the unwanted transmission. You're most likely to receive interference from devices in your own home or close by because the signals are weak. If you're sure that the source isn't on your property, you need a portable receiver that can hear the interfering signal.

The ARRL RFI website has some helpful hints on each type of interference, as well as guidance on how to diplomatically address the problem (because it's not your device). The webOverview page of the ARRL RFI website contains excellent material on dealing with and managing interference complaints (both by you and from others). ARRL members have access to the league's technical coordinators and technical information services.

Ferrites as RFI suppressors

Ferrite is a magnetic ceramic material that's used as a core in RF inductors and transformers. It's formed into rods, *toroid cores* (circular and rectangular rings), and beads (small toroids made to slip over wires). Ferrite has good magnetic characteristics at RF and is made in different formulations, called *mixes,* that optimize it for different frequency ranges. Ferrites made of Type 31 material work best at HF, for example, and Type 43 ferrites work best at VHF.

Winding a cable or wire on a ferrite core or rod creates *impedance* (opposition to AC current) that tends to block RF signals. The more turns on the core, the higher the impedance. Because they're small, you can place ferrite cores very close to the point at which an undesired signal is getting into or out of a piece of equipment. You can secure cores on the cable with a plastic cable tie, tape, or heat-shrink tubing. This technique works particularly well with telephone and power cords. *Split cores* come with a plastic cover that holds the core together, which makes placing the core on a cable or winding turns easy if the cable already has a large connector installed.

If you want to find out more about ferrites, the online tutorial "A Ham's Guide to RFI, Ferrites, Baluns, and Audio Interfacing," written by Jim Brown (K9YC), is very good; you can find it at `www.audiosystemsgroup.com/RFI-Ham.pdf`. The *ARRL Handbook* and *The ARRL RFI Book* also go into great detail about how to use ferrites to fight noise and interference.

You can eliminate or reduce most types of interference to insignificant levels with careful investigative work and application of the proper interference-suppression techniques. The important thing is to keep frustration in check and work the problem through.

Building Equipment from a Kit

Building your own gear — even just a simple speaker switch — is a great ham tradition. By putting equipment together yourself, you become familiar with the operation, repair, and maintenance of your existing equipment.

If you're just getting started in electronics, I recommend that you start your building adventures with kits. When I got started, you could find the Heathkit label on equipment in every ham shack. Today, kits are available from many sources, such as Ramsey, Ten-Tec, Vectronics, RadioShack, and others. For an up-to-date list of companies selling ham equipment kits, check out the list of vendors at `www.ac6v.com/kits.htm`.

Choose simple kits until you're confident about your technique. Kits are great budget-saving ways to add test instruments to your workbench and various gadgets to your radio station. Also, you don't have to do the metalwork, and the finished result looks great.

After you build a few kits, you'll be ready to move up to building a complete radio. Although the Elecraft K3 (`www.elecraft.com`) is the top-of-the-line radio kit available today, numerous smaller QRP radio kits are available from other vendors.

You can build most kits by using just the maintenance tool kit described at the beginning of this chapter. Concentrate on advancing your soldering skills. Strive to make the completed kit look like a master built it, and take pride in the quality of your work. Read the manual and use the schematic to understand how the kit works. Observe how the kit is put together mechanically, particularly the front-panel displays and controls.

Building Equipment from Scratch

Building something by starting with a blank piece of paper or a magazine article and then putting it to use in your own station is a real thrill. Building from scratch isn't too different from building from a kit, except that you have to make your own kit. Your first project should be a copy of a circuit in a magazine or handbook — one that's known to work and that comes with assembly and test directions. If a blank printed circuit board is available, I recommend ordering one.

Imagine that you have to make a kit for someone else based on the instructions, schematic, and list of components. Photocopy the article, and highlight all the instructions. If an assembly drawing is included, enlarge it for guidance. Make extra copies so you can mark them up as you go. Read the article carefully to identify any critical steps. When you get your components together, sort them by type and value, and place them in jars or the cups of an old muffin pan. Keep a notebook handy so that you can take notes for later use. As you build and test the unit and finally put it to use, everything is completely documented.

If you choose to design a circuit from scratch, I salute you! Documenting your work in a notebook is even more important for a project that starts with design. Take care to make your schematics complete and well-labeled. Record whatever calculations you must make so that if you have to revisit some part of the design later, you have a record of how you arrived at the original values. Digital cameras are abundant, so take a few photos at important milestones of construction. When you finish, record any tests that you make to verify that the equipment works.

Don't let failure get you down! First designs hardly ever work out exactly right, and sometimes, you even wind up letting all the smoke out of a component or two. If a design doesn't work, figure out why and then move on to the next version. Don't be afraid to ask for help or to try a different angle. Ham radio isn't a job, so keep things fun. After all, it's *amateur* radio!

Part V
The Part of Tens

Enjoy an additional Part of Tens chapter about ways to give back to ham radio. It's online at `www.dummies.com/extras/hamradio`.

In this part . . .

- Benefit from ten simple but valuable lessons for beginners.
- Follow up with ten tips that the masters use on the air every day.
- Find out how to make your first station work effectively for you.
- Fun is an important part of ham radio; here are some tips to help you have plenty!
- Enjoy an additional Part of Tens chapter about ways to give back to ham radio. It's online at `www.dummies.com/extras/hamradio`.

Chapter 16

Ten Secrets for Beginners

In This Chapter

- Timeless tips for improvement and enjoyment
- Ways to expand your proficiencies and build confidence

In this chapter, I present ten fundamental truths that can help even the rankest beginner keep the wheels turning during those first forays into ham radio. Keep these tips in mind, and you'll be on your way to veteran status in no time.

Listen, Listen, Listen

Listening is the most powerful and important way to learn. Listen to the successful stations (or watch them, if you're using a digital mode) to study their techniques. Listening to on-the-air contacts is called *reading the mail.*

All ham communications are open and public; they can't be encrypted or obscured. Just turn on the radio and get a real-time seminar in any facet of ham radio communication techniques you care to try.

Buddy Up

Find a friend who's learning the ropes, the way you are. Is someone from your licensing class or club also getting started? Meet on the air, and get used to using your equipment together. Attend club meetings and functions together. The best part is sharing in each other's successes!

Know Your Equipment

Hams joke about never reading the owner's manual, but don't believe it. Hams need to know their equipment. Here are a few pointers:

- Look for online videos showing how to use or operate the equipment. If a demo or tutorial is available to you, go through it.
- Many models of transceivers have their own user groups, such as those on Yahoo! Groups or Google Groups. These groups can be great resources.
- Find a club member with the same radio. Most likely, he or she will be glad to let you have a closer look or explain how it works.
- Practice adjusting the main controls or settings to observe the effects.
- Acquire at least passing familiarity with even the most obscure controls.
- Keep the manual handy for quick reference, too.

Follow the Manufacturer's Recommendations

Manufacturers want you to get the best performance and satisfaction out of their equipment, don't they? That's why they have recommended settings and procedures. Follow the recommendations until you're comfortable enough to optimize performance on your own.

Do what the manufacturer says — at least until you become an expert.

Try Different Things

Don't feel that you have to stay with one mode, band, website, radio, or club. Changing your mind and striking out in a different direction is okay. As you become more comfortable with ham radio, feel free to dabble in anything that catches your fancy. Sooner or later, you'll discover something that makes you want to dive in deeply.

Know That Nobody Knows Everything

Surround yourself with handbooks and how-to articles, magazines, websites, manuals, and catalogs. Use any reference that's available. If you're confused or not getting the results you expect, ask someone at a club meeting, on the air, or in an Internet forum for help.

The oldest tradition in ham radio is hams giving other hams a hand. We're all amateurs. We like to do what we do. Someone helped us, and we'll help you.

Practice Courtesy

Behind every receiver is a person just like you. Polite terms like "Please," "Thanks," "Excuse me," and "Sorry" work just as well on the air as they do in person. Listen before transmitting, and be flexible.

If you encounter a rude operator, just change frequencies or bands, or find something else to do. Don't let your temper escalate on the air.

Join In

By definition, ham radio isn't a solitary activity. It's a lot more fun if you have some regular acquaintances. Being welcomed on the air into a roundtable QSO or a local club net is great. Ham radio welcomes kings and paupers equally, and we're all on a first-name basis. The ham bands are your home. See Chapter 3 for some tips on finding and joining ham radio events in person and on the air.

Get Right Back in the Saddle

So what if you called CQ and nobody responded? So what if you put up a new antenna and it didn't work? Get right back in the saddle, and try again. I don't know of any ham who had instant success right off the bat, so don't get discouraged. You worked too hard to get that license to give up.

Relax, It's a Hobby!

I know the scary feeling of thinking that every ham is listening whenever you get on the air. Hey, relax. Don't worry that a mistake is going to put you on a hobbywide blacklist. If you try something new and it doesn't work out, that's okay. Everybody fumbles now and then. Keep ham radio fun for yourself, and do things you enjoy.

Chapter 17

Ten Secrets of the Masters

In This Chapter

- Time-tested know-how
- Methods of honing your expertise

You may think that grizzled ham radio veterans surely have stores of secret knowledge that took years and years to acquire — knowledge that makes them the masters of all they survey. Certainly, the veterans have experience and expertise, but they also rely on simple principles that work in many situations. You can use these principles, too.

Listening to Everything

Masters get more out of listening than anyone else because they've learned how to listen. Every minute you spend listening is a minute learning and a minute closer to being a master.

Looking under the Hood

Operating a radio and building an efficient, effective station are much easier if you know how the equipment works. Even if you're not terribly tech-savvy, take the time to get familiar with the basics of electronics and how your equipment functions. A master understands the effects of controls and their adjustments.

Studying Ham Radio History

An appreciation of the rich history of ham radio helps you understand how the hobby has shaped itself. Ham radio is full of conventions and methods developed over time, many of which seem to be confusing and obscure at first. You can master these conventions by understanding how they came about, which will help you adapt quickly to new situations and techniques.

Keeping the Axe Sharp

When asked what he would do if he had eight hours to cut down a tree, Abe Lincoln replied that he would spend the first six hours sharpening his axe. Masters keep their equipment and skills sharp. When they're needed on the air, they're ready.

Practicing to Make Perfect

Even a sharp axe, however, gets dull if it isn't used. A master is on the air regularly, keeping in touch with conditions. A master knows what stations are active, from where, and when, as well as when important nets and on-the-air events take place.

Make operating your radio station a natural and comfortable activity by keeping yourself in shape with regular radio exercise.

Paying Attention to Detail

Masters know that the little things are what make the difference between 100 percent and 90 percent performance — or even between being on the air and off the air. The most expensive station isn't worth a nickel if it doesn't work properly when you need it. Waterproofing that connector completely or having your CQ sound just right really pays off in the long run. Masters are on the radio for the long run.

Knowing What You Don't Know

Take a tip from Mark Twain, who said, "It's what you know for sure that just ain't so." If you get something wrong, don't be too proud to admit it. Find out the right way; track down the correct fact. People make their worst mistakes by ignoring the truth.

Radio waves and electricity don't care about human pride. A master isn't afraid to say three dreaded words: "I don't know."

Taking Great Care of Antennas

If you look at the top stations in any facet of ham radio, you'll find that the owners spend the most time and effort on their antenna systems. You'll find no better return on investment in ham radio than in improving your antennas. Masters are often antenna gurus.

Respecting Small Improvements

Masters know that any improvement in the path between operators is not to be discounted. Anything that makes your signal easier to understand — 1 dB (decibel) less noise received, 1 dB better audio quality, 1 dB stronger transmitted signal — makes the contact easier.

Being a Lifelong Student

Your ham radio license is really a license to study. Take advantage of every learning opportunity, including learning from your mistakes. (You'll have plenty!) Each problem or goof is also a lesson.

Masters got to be masters by starting as raw recruits just like you and then making one improvement at a time, day in and day out.

Chapter 18

Ten First-Station Tips

In This Chapter

- Avoiding mistakes in putting together your first station
- Saving money on equipment

When you're putting your first station together — whether it's a home, mobile, or portable station, or even just a handheld radio — getting sidetracked in a way that creates problems later is easy. Avoid common pitfalls by applying the simple tips I provide in this chapter. That way, you won't scratch your head later, wondering "Why did I do that?"

Be Flexible

Don't assume that you'll be doing the same activities on the air forever. Here are a few tips on flexibility:

- Avoid using overspecialized gear except where it's required for a specific type of operating.
- Use a computer and software to implement functions that are likely to change, such as operating on the digital modes.
- Don't nail everything down physically. Allow equipment to be moved around for comfort and layout convenience. The built-in look is attractive but very hard to change if you change your mind later.

Study Other Stations

Browse the web for articles and videos that show how other stations are put together and operated. Make note of any particularly good ideas. Don't be intimidated by big stations, because they started out as small stations!

Don't hesitate to contact the station owners with questions; they welcome your attention and interest. Take advantage of opportunities to visit local stations, too.

Budget for Surprises

Spending a lot of money on a radio right away is tempting, but you'll find yourself needing other gear — such as antennas and cables — that perhaps you hadn't counted on. Those extras can add up to at least as much as your main radio, so leave yourself some budget for them.

Shop for Used-Equipment Bargains

If you have a knowledgeable friend who can help you avoid worn-out and inadequate gear, buying used equipment is a great way to get started. Purchasing used gear from a dealer who offers a warranty is also a good option. Saving money now leaves you more cash for exploring new modes and bands later.

Caveat emptor: You can easily encounter obsolete or poorly functioning equipment when you're shopping for used gear. If you're in doubt, if you can't check it out, or if the deal seems too good to be true, pass it up.

Build Something Yourself

Using equipment that you build yourself is a thrill. Start small by building accessory projects such as audio switches, filters, and keyers. Building things yourself can save you some money, too. Don't be afraid to get out the drill and soldering iron. You can find lots of kits, web articles, magazines, and books of projects to get you started.

Get Grounded Early

Don't neglect grounding and bonding. Put in the ground system (see Chapter 13) as the first step; adding grounding after the equipment and wires are already in place usually is more trouble.

Keeping equipment connected properly also helps prevent feedback and ground loops, both of which are frustrating and aggravating problems.

Save Cash by Building Your Own Cables

You need lots of cables and connectors in your station. At a cost of roughly $5 or more for each premade cable, you can quickly spend as much on connecting your equipment as you can on purchasing a major accessory. Learn how to install your own connectors on cables, and you'll save many, many dollars over the course of your ham career.

Build Step by Step

After you have the basics of your shack in place, upgrade your equipment in steps so that you can always hear a little farther than you can transmit. Don't be an alligator (all mouth, no ears). Plan with a goal in mind so that your ham radio dollars and hours all work to further that goal.

Find the Weakest Link

Every station has a weak link. Always be on the lookout for a probable point of failure or of loss of quality. On the airwaves, you'll encounter stations with a multibucks radio but a cheap, garage-sale microphone that results in muffled or distorted audio. Use quality gear, and keep heavily used equipment well maintained.

Make Yourself Comfortable

You're going to spend a lot of hours in front of your radio, so take care of yourself, too. Start with a comfortable chair. Excellent chairs are often available in used-office-furniture stores at substantial discounts. Also make sure that you have adequate lighting and that the operating desk is at a comfortable height. The dollars you spend will pay dividends every time you go on the air.

Chapter 19

Ten Easy Ways to Have Fun on the Radio

In This Chapter

- Tips to break out of your routine
- Contests and events

So you're sitting there saying, "There's nothing to do!" Don't worry. Everybody gets in a bit of a rut now and then. In this chapter, I give you ten great ideas to shake off the radio blues and spice up your operating.

Find People Having Fun — and Join In

Sounds simple, doesn't it? Just turn on the radio, and start tuning the bands. Join a ragchew, check into a net, monitor a slow-scan transmission, listen for someone calling "CQ contest," or find a pileup and dive in. On any given day, hundreds of activities are taking place; all you need to do to participate is spin the tuning dial and pay attention.

Take Part in Special Events and Contests

Every weekend, you can find little pockets of activity sprinkled around the HF and VHF bands as contests and special-event stations take to the airwaves. Nearly every contest is open to casual passersby. The operators involved welcome your call and help you exchange the necessary information.

Some special-event stations send beautiful, interesting certificates for contacts.

Challenge Yourself

Get together with your club friends, and dream up a challenge for the other members. Who can contact the most states in a weekend, for example, or work a special-event station on the most bands and modes? Make up a funny name for the exercise. Challenge another club to a team competition in an upcoming contest. Buy an old trophy and award it to the winner at the next club meeting. Go nuts; it's a hobby!

Collect Some Wallpaper

Hams love awards, and thousands of awards are available. Find a simple one, such as the Worked All Continents award (`www.iaru.org/worked-all-continents-award.html`), and start tuning. Or find stations with the same call-sign suffix as yours, exchange QSL cards (see Chapter 14), and start hanging the cards on the wall of your shack. Watch out; you won't be able to work just one!

Join the Parade

Most larger cities and towns sponsor events such as parades and festivals, . Your club might assist the organizers by handling communications and other electronics chores. You'll feel great after you give these folks a hand, and they'll come to appreciate ham radio, too.

Contact your ARRL section manager for information about how to get in touch with organizers, or get in touch with your local ARES team. See Chapter 3 for details about both organizations.

Go Somewhere Cool

For a real treat, try mobile or portable operation from some unusual place. On HF, try operating from a rural county or county line on the County Hunters Net. On VHF and UHF, a short drive can put you in a sought-after grid square (see Chapter 11). The nearest hilltop or scenic overlook can generate hours of fun. Take a photo, and make up a neat QSL card (see Chapter 14) to go with the contacts.

Squirt a Bird

Making contacts through a satellite — a process called *squirting a bird* — is a lot easier than you think. Try the FM repeater satellites for starters, as described in Chapter 11 and on the AMSAT website (`http://ww2.amsat.org`). Monitor the downlink frequencies of the International Space Station (ISS). You'll receive an interesting certificate from the ARISS program (`www.arrl.org/amateur-radio-on-the-international-space-station`) just for reporting that you heard the ISS transmissions.

Watching a satellite move across the sky at dusk or dawn will never be the same after you make contacts through one.

Pick Up a Second (Or Third) Language

Brush up those foreign-language skills, and make contact with a DX (far away) station. If you're equipped for HF, you can talk to a DX station directly. Chapter 9 explains how to make VHF/UHF contacts via repeaters by using the IRLP, EchoLink, or D-STAR linking system.

Ham radio gives you many new language opportunities. You can start small with "Hello," "Goodbye," and "Thank you." Work your way up from there!

Do Shortwave Listening

What's on the frequencies *between* the HF ham bands? Most radios now have general-coverage receivers that tune all frequencies, so why not tune in England's national broadcast system, the BBC, as it broadcasts to countries around the world? Also, China Radio International broadcasts a strong signal to North America on several frequencies. Check the station websites or get a copy of a shortwave listener's guide to find English-language programs.

Music is always available and requires no translation to enjoy.

Visit a New Group

Drop in on a new club or net any time. The members will be pleased that you're visiting, and you may gain a whole new outlook on the hobby. Be sure to invite your new friends to your club's meetings in return.

I particularly enjoy finding a club program on a subject I'm not acquainted with — a great way to discover something new and also make a friend or two.

Appendix A

Glossary

This glossary includes material contributed by the ARRL. A broader glossary is available at `www.arrl.org/ham-radio-glossary`.

AC: See *alternating current (AC)*.

AC hum: Unwanted 60 Hz or 120 Hz modulation of an audio RF signal due to inadequate filtering in a power supply or improper grounding. See also *radio frequency (RF)*.

AC voltage: A voltage with a polarity that reverses at regular intervals.

active filter: See *filter*.

adapters: Connectors that convert one type of connector to another.

airband: The frequencies from 108 MHz to 137 MHz, used for civil aviation communications and navigation.

allocations: Frequencies authorized for a particular FCC radio communications service.

alternating current (AC): Electrical current with a direction that reverses at regular intervals.

AM band: The commercial AM broadcast band, spanning 530 kHz to 1710 kHz.

amateur operator: A person who has written authorization to be the control operator of an amateur station.

Amateur Radio Emergency Service (ARES): A group, sponsored by the ARRL, that provides emergency communications. ARES works with groups such as the American Red Cross and local emergency operations centers.

amateur service: A radio communication service for the purpose of self-training, intercommunication, and technical investigations carried out by amateurs.

amateur station: A station licensed to make transmissions in the amateur service.

amateur television (ATV): Analog fast-scan television using commercial transmission standards (NTSC in North America).

American Radio Relay League (ARRL): The national association for amateur radio (`www.arrl.org`).

ammeter: A test instrument that measures current.

antenna: A structure designed to radiate and receive radio waves.

beam: See *directional antenna.*

CW: Continuous wave, usually used to refer to Morse code transmissions.

data modes: Communications in which the information is transmitted as individual digital characters.

deci (or d): The metric prefix for 10^{-1} or division by 10.

degree: A measure of angle or phase. There are 360 degrees in a circle or a cycle, for example.

delta loop antenna: A loop antenna in the shape of a triangle.

demodulate: The process by which the information in a modulated signal is recovered in its original form.

designator: Letters and numbers used to identify a specific electronic component.

detect: (1) To determine the presence of a signal. (2) To recover the information directly from a modulated signal.

detector: The stage in a receiver in which the modulation (voice or other information) is recovered from a modulated RF signal.

deviation: The change in frequency of an FM carrier due to a modulating signal; also called *carrier deviation.*

dielectric: The insulating material in which a capacitor stores electrical energy.

diffract: To alter the direction of a radio wave as it passes by edges of or through openings in obstructions such as buildings or hills. *Knife-edge diffraction* results if the dimensions of the edge are small in terms of the wave's wavelength.

digipeater: A type of repeater station that retransmits or forwards digital messages.

digital mode: See *data modes*.

digital signal: (1) A signal (usually electrical) that can have only certain specific amplitude values or steps, usually two: 0 and 1 or ON and OFF. (2) On the air, the same as a data mode signal. See also *data modes*.

digital signal processing (DSP): The process of converting an analog signal to digital form and using software to process the signal in some way, such as filtering or reducing noise.

diode: An electronic component that allows electric current to flow in only one direction.

diplexer: A device that allows radios on two different bands to share an antenna. Diplexers are used to allow a dual-band radio to use a single dual-band antenna.

dipole antenna: A popular antenna consisting of a length of wire or tubing usually half a wavelength long, with its feed point in the middle.

direct conversion: A type of receiver that recovers the modulating signal directly from the modulated RF signal.

direct current (DC): Electrical current that flows in only one direction.

direct detection: The wiring of a device acting as an unintentional receiver by converting a strong RF signal directly to voltages and currents internally, usually resulting in RF interference to the receiving device.

directional antenna: An antenna with enhanced capability to receive and transmit in one or more directions.

directional wattmeter: See *wattmeter*.

director: A parasitic element of a directional antenna that focuses the radiated signal in the forward direction.

discharge: To extract energy from a battery or cell. *Self-discharge* refers to the internal loss of energy without an external circuit.

downlink: Transmitted signals or the range of frequencies for transmissions from a satellite to Earth.

Federal Communications Commission (FCC): The government agency that regulates telecommunications in the United States.

feed line: See *transmission line*.

filter: A circuit designed to manipulate signals differently on the basis of their frequency. Typical filters used by amateurs include *low-pass, high-pass,* and *bandpass* functions.

FM band: The commercial FM broadcast band, spanning 88 MHz to 108 MHz.

giga (or G): The metric prefix for 10^9 or multiplication by 1,000,000,000.

go kit: A prepackaged collection of equipment or supplies kept on hand to prepare an operator to respond quickly in time of need.

grace period: The time allowed by the FCC following the expiration of an amateur license to renew that license without taking an examination. An operator who holds an expired license may not operate an amateur station until the license is reinstated.

grant: Authorization given by the FCC.

grid square: A locator in the Maidenhead Locator System that divides the surface of the Earth into rectangles 1 degree of latitude by 2 degrees of longitude.

ground: (1) The electrical potential of the Earth's surface; also called *Earth ground* or *ground potential.* (2) A local common voltage reference for a circuit.

ground loss: Power converted to heat as a radio wave reflects from or travels along the Earth's surface or through the ground.

ground plane: A conducting surface of continuous metal or discrete wires that creates an electrical image of an antenna. Ground-plane antennas require a ground plane to operate properly.

ground rod: A copper or copper-clad steel rod driven into the Earth to create a ground connection.

ground-wave propagation: Propagation in which radio waves travel along the Earth's surface.

half-wave dipole: See *dipole antenna.*

ham band receiver: A receiver designed to receive only frequencies in the amateur bands.

hamfest: A flea market for ham radio, electronic, and computer equipment and accessories.

handheld radio: A VHF or UHF transceiver that can be carried in the hand or pocket.

harmful interference: Interference that seriously degrades, obstructs, or interrupts a radio communication service operating in accordance with the FCC's radio regulations.

harmonic: A signal at an integer multiple (2, 3, 4, and so on) of a lowest or fundamental frequency.

header: The first part of a digital message, containing routing and control information about the message. See also *preamble*.

headset (or boomset): A set of headphones with a microphone mounted on it for additional convenience.

health-and-welfare traffic: Messages about the well-being of people in a disaster area. Such messages must wait for emergency and priority traffic to clear and result in advisories to those outside the disaster area awaiting news of family members and friends.

henry (H): The basic unit of inductance.

high frequency (HF): The frequency range of 3 MHz to 30 MHz.

load: (1) A device or system to which electrical power is delivered, such as a heating element or antenna. (2) The amount of power consumed, such as a 50-watt load.

lobe: A direction of enhanced reception or transmission in an antenna's radiation pattern. The *main lobe* is the lobe with the greatest strength for the entire pattern. A *side lobe* is directed at an angle to the main lobe.

log: The documents of a station that detail its operation. A log provides a record of who was acting as a station's control operator and information about troubleshooting interference-related problems or complaints.

loop: An antenna with element(s) constructed as continuous lengths of wire or tubing.

lower sideband (LSB): (1) In an AM signal, the sideband located below the carrier frequency. (2) The common single sideband operating mode on the 40, 80, and 160 meter amateur bands.

lowest usable frequency (LUF): The lowest-frequency signal that can be returned to Earth by the ionosphere without being absorbed.

low-pass filter (LPF): A filter designed to pass signals below a specified cutoff frequency while attenuating higher-frequency signals.

machine: Slang for *repeater.*

maximum usable frequency (MUF): The highest-frequency radio signal that can reach a particular destination via sky-wave propagation. See also *sky-wave.*

mayday: From the French *m'aidez* (help me); used when calling for emergency assistance in voice modes.

mega (M): The metric prefix for 10^6 or multiplication by 1,000,000.

memory channel: Frequency and mode information stored by a radio and referenced by a number or alphanumeric designator.

meteor scatter: Communication by signals reflected by the ionized trails of meteors in the upper atmosphere.

meter: (1) A reference to the wavelength of a signal, usually used to indicate a specific amateur or shortwave broadcasting band. (2) A device that displays a numeric value as a number or as the position of an indicator on a numeric scale.

micro (μ): The metric prefix for 10^{-6} or division by 1,000,000.

microphone: A device that converts sound waves to electrical energy (abbreviated *mic* or *mike).*

microwave: A conventional way of referring to radio waves or signals with frequencies greater than 1000 MHz (1 GHz).

modulate: The process of adding information to a carrier signal so that it may be transmitted.

potentiometer (pot): See *variable resistor.*

preamble: The initial portion of a digital message preceding the data.

radio frequency (RF): Signals with a frequency greater than 20 kHz, the upper end of the audio range.

repeater: A station that receives signals on one frequency and simultaneously retransmits them on another frequency to relay them over a wide range.

shielding: Surrounding an electronic circuit with conductive material to block RF signals from being radiated or received.

short circuit: (1) An electrical connection that causes current to bypass the intended path. (2) An accidental connection that results in improper operation of equipment or circuits.

sideband: An RF signal that results from modulating the amplitude or frequency of a carrier. An AM sideband can be either higher in frequency (upper sideband, or USB) or lower in frequency (lower sideband, or LSB) than the carrier. FM sidebands are produced on both sides of the carrier frequency.

signal generator: A device that produces a low-level signal that can be set to a desired frequency.

signal report: An evaluation of the transmitting station's signal and reception quality.

simplex: A method of operation that involves receiving and transmitting on the same frequency.

sine wave: A waveform with an amplitude proportional to the sine of frequency × time.

single sideband (SSB): A form of amplitude modulation in which one sideband and the carrier are removed.

skip: See *sky-wave*.

skip zone: An area of poor radio communication, too distant for ground-wave propagation and too close for sky-wave propagation.

skyhook: Slang for *antenna* or *antenna support.*

sky-wave: The method of propagation by which radio waves travel through the ionosphere and back to Earth; also referred to as *skip.* Travel from the Earth's surface to the ionosphere and back is called a *hop.*

slow-scan television (SSTV): A television system used by amateurs to transmit pictures within the bandwidth required for a voice signal.

SOS: A Morse code call for emergency assistance.

space station: An amateur station located more than 31.1 miles (50 km) above the Earth's surface.

speaker: A device that turns an audio frequency's electrical signal into sound.

spectrum: The range of electromagnetic signals. The radio spectrum includes signals between audio frequencies and infrared light.

speech compression or processing: Increasing the average power of a voice signal by amplifying low-level components of the signal more than high-level components.

splatter: A type of interference with stations on nearby frequencies that occurs when a transmitter is overmodulated.

transmission line: Two or more conductors arranged so as to work together in carrying AC electrical energy from one point to another.

ultra high frequency (UHF): The frequency range between 300 MHz and 3 GHz.

Universal Licensing System (ULS): FCC database for all FCC radio services and licensees.

uplink: Transmitted signals or the range of frequencies for transmissions from Earth to a satellite. See also *downlink.*

upper sideband (USB): (1) In an AM signal, the sideband located above the carrier frequency. (2) The common voice single sideband operating mode on the 60, 20, 17, 15, 12, and 10 meter HF amateur bands, and all the VHF and UHF bands.

vacuum tube: An electronic component that operates by controlling electron flow between two or more electrodes in a vacuum.

vanity call: A call sign selected by the amateur instead of one sequentially assigned by the FCC.

variable resistor: A resistor whose value can be adjusted, usually without removing it from a circuit.

variable-frequency oscillator (VFO): An oscillator with an adjustable frequency. A VFO is used in receivers and transmitters to control the operating frequency.

vertical antenna: A common amateur antenna with a vertical radiating element.

very high frequency (VHF): The frequency range between 30 MHz and 300 MHz.

visible horizon: The most distant point a person can see by line of sight.

voice communications: Any of several methods used by amateurs to transmit speech. Hams can use several voice modes, including FM and SSB.

voice-operated transmission (VOX): Turning a transmitter on and off under control of the operator's voice.

volt (V): The basic unit of electrical potential or electromotive force.

voltage: The electromotive force or difference in electrical potential that causes electrons to move through an electrical circuit.

voltmeter: A test instrument used to measure voltage.

Volunteer Examiner (VE): A licensed amateur who is accredited to administer amateur license examinations.

Volunteer Examiner Coordinator (VEC): An organization that has entered into an agreement with the FCC to coordinate amateur license examinations.

waterfall display: Used with digital modes, a display that consists of a sequence of horizontal lines showing signal strength as a change of brightness and color, with the signal's frequency represented by position on the line. Older lines move down the display so that the history of the signal's strength and frequency form a waterfall-like picture.

watt (W): The unit of power in the metric system.

wattmeter: A test instrument used to measure the power output (in watts) of a transmitter. A *directional wattmeter* measures both forward and reflected power in a feed line.

waveform: The amplitude of an AC signal as it changes with time.

wavelength: The distance a radio wave travels during one cycle. The wavelength relates to frequency in that higher-frequency waves have shorter wavelengths.

Appendix B

The Best References

This appendix lists useful resources that can answer your many questions as you begin to experience and enjoy ham radio. Most of the books are available from the ARRL Store (`www.arrl.org/shop`) and CQ Communications (`www.cq-amateur-radio.com`), and some are available in e-book format. When a resource has a special publisher or other source, I list a web address where you can find it.

Specific tools, such as software apps and models of equipment, are discussed throughout the book where the associated topics are discussed. These tools are likely to change versions and model numbers rapidly, because more are created every day. Use the resources listed in this appendix and your favorite Internet search engine to keep track of what is currently available or supported.

Club and special-interest group newsletters, user groups' websites, and e-mail reflectors are also excellent sources of information; many are archived and searchable, so you can look up what you need to know.

Finally, don't forget your mentor or Elmer — the best resource of all!

Web Portals

The web portals referenced here are good "newsstands" for ham radio. On these sites, you'll find information about current and upcoming events, radio conditions, and news stories. These sites also host mailing lists and forums on various topics, equipment swap-and-shops, archived files and photos, and product reviews. Each site is designed to be your ham radio home page.

- **AC6V's Amateur Radio & DX Reference Guide** (`www.ac6v.com`)**:** This site offers many links and references covering all phases of ham radio.
- **The DXZone** (`www.dxzone.com`)**:** The DXZone is another site that collects links, references, and articles about practically any ham radio topic you can think of.

- **eHam.net** (www.eham.net): This portal features news and articles, e-mail discussion groups, product reviews, for-sale listings, DX spotting and solar information, and surveys, among other things.
- **QRZ.com** (www.qrz.com): QRZ.com is a general-interest portal and call sign/licensee lookup facility that features extensive articles, news and discussion groups, and online practice licensing exams.
- **QSL.net** (www.qsl.net): QSL.net is host to hundreds of individual and club ham radio web pages and e-mail reflectors. You can search the links and pages for topics of interest.
- **Yahoo! Groups and Google Groups** (http://groups.yahoo.com and http://groups.google.com): These sites offer links to a large number of general-purpose and specialized ham radio interest groups and clubs. Both are great resources for finding local clubs and getting advice from knowledgeable hams.

Online Programs

Like everybody else, hams have taken advantage of online media. Following are some of the popular ham programs that were being produced and distributed in early 2013. The web portals listed in the preceding section are good places to find new programs.

The most widely viewed programs are in interview and talk-show style. The topics cover everything from emergency communications to awards to news of on-the-air events. How-to segments on antennas or electronics are frequently part of the mix.

- **AmateurLogic.TV** (www.amateurlogic.tv): General-interest programs about ham radio and radio technology.
- **Amateur Radio Newsline** (www.arnewsline.org): News stories about current events in ham radio.
- **Ham Nation** (http://twit.tv.hn): Talk show–style programs about ham radio with regular and invited guests.
- **Ham Radio Now** (http://arvideonews.com/hrn): Interviews with individual hams in the news, product reviews, and tutorials.
- **SolderSmoke** (www.soldersmoke.blogspot.com): Emphasizing technical programs, SolderSmoke is an audio podcast discussing anything that involves radio technology. It's very hands-on and fun for do-it-yourselfers, hams or not.

- **WWROF webinars** (`http://wwrof.org`)**:** The World Wide Radio Operators Foundation sponsors and archives webinars on a variety of topics that mostly have to do with station-building, propagation, and contest operating — high-performance radio.

Tutorials and demonstrations abound on video services such as YouTube (`www.youtube.com`) and Vimeo (`http://vimeo.com`). The ARRL's *Digital QST* magazine (see "Magazines," later in this appendix) also includes numerous interesting videos, such as live demonstrations of new radio equipment.

Operating References

No one knows or remembers everything, so it's a good idea to have references on hand to guide your on-the-air activities. Here are some books you'll find handy on a day-to-day basis. All books except the last one in this list are available at the ARRL Store (`www.arrl.org/shop`).

- ***ARRL Operating Manual:*** This operating guide, published by the ARRL, covers nearly all phases of ham radio operating, including maps and numerous references.
- ***ARRL Repeater Directory:*** Published by the ARRL, this directory lists North American repeaters on 10 meters through the UHF and microwave bands.
- ***FCC Rules and Regulations for the Amateur Radio Service:*** The rulebook explains in clear text what the regulations mean and how to apply them. It also includes the actual text of the Part 97 rules.
- ***Getting Started with Ham Radio:*** This book, by Steve Ford (WB8IMY), is a good introductory text to help you set up a station and get on the air.
- ***Two-Way Radios and Scanners For Dummies:*** My 2005 book (John Wiley & Sons, Inc.) discusses many other communication services that complement ham radio nicely.

Public service

After you start performing public-service activities, you'll need forms and training. Luckily, most of what you need is available at the following sites.

- **Federal Emergency Management Agency** (`www.fema.gov`)**:** FEMA, the national agency for emergency and disaster response, provides comprehensive information on personal and community preparedness as well as programs that support local and regional public safety agencies.

- **FEMA Emergency Response Training (**`www.fema.gov/training-0`**):** If you volunteer for an ARES team (see the next item), you need to be familiar with the training materials provided by FEMA and probably have to pass a few of the introductory certification courses.
- **ARES Emergency Communications Training (**`www.arrl.org/emergency-communications-training`**):** This collection of online courses and training resources introduces you to emcomm, net control and management, and team management and planning.
- **ARRL Online Net Directory Search (**`www.arrl.org/arrl-net-directory-search`**):** You can use ARRL's Net Directory Search to find on-the-air nets by type, topic, frequency, name, or region.
- **ARRL Public Service (**`www.arrl.org/public-service`**):** This web page includes several operating manuals available for free download, guidelines, brochures, and links to other emergency communications organizations.

Digital or data modes

The information available at these sites really helps you get going on the digital modes. The two modes listed here — PSK and RTTY — are the most popular for new hams.

- **AA5AU RTTY Page (**`www.aa5au.com/rtty.html`**):** There's no better place to start learning about RTTY operation than Don Hill (AA5AU)'s website, especially if you'd like to give RTTY contest operation a try.
- **PODXS Ø7Ø Club (**`www.podxs070.com`**):** The PODXS (Penn-Ohio DX Society) Ø7Ø Club website has everything a prospective PSKer could want, including FAQs, a help desk, contests for PSK users, and an awards program. A club reflector is available at `http://groups.yahoo.com/group/070`.
- **PSK31 (**`www.qsl.net/wm2u/psk31.html`**):** Ernie Mills (WM2U) has assembled a terrific website that shows you how to get on the air with PSK31. The site features lots of links to programs that support PSK31 and provide additional information.
- **Tucson Amateur Packet Radio (**`www.tapr.org`**):** TAPR is the biggest ham group specializing in digital modes. Its website is a smorgasbord of information about all the popular digital protocols, focusing on the technical details.

DXing resources

Successful DXing calls for timely information about which stations are active and what conditions you can expect on the air. The following references are DXing magazines, newsletters, and bulletins containing information about active and upcoming DX events.

- ***The Complete DX'er*** **(**`www.idiompress.com/books-complete-dxer.html`**):** Every budding DXer should read this book, by Bob Locher (W9KNI), to understand the basics of DXing and good operating. In wonderfully readable short-story format, Locher explains the right ways to go about putting DX call signs in your log. If you're mystified by what you hear while chasing DX, Locher can help you understand the whys and hows.
- ***The Daily DX*** **(**`www.dailydx.com`**):** *The Daily DX* is a daily e-mail newsletter full of late-breaking news and on-the-air surprises.
- ***DX Magazine*** **(**`www.dxpub.net/DX-Magazine.html`**):** This bimonthly periodical contains articles about the techniques of DXing and includes numerous travelogues of DXpeditions to interesting places around the world.
- **DX Summit (**`www.dxsummit.fi`**):** DX Summit, developed and operated by the Finnish group Radio Arcala, is a worldwide DX spotting website where hams post reports on every band from every land 24 hours a day. The reports, known as *spots,* are searchable by call and band. The site also records solar data for propagation information.
- ***World Radio and TV Handbook*** **(**`www.wrth.com`**):** This reference guide is for shortwave broadcast listeners. It's available online, on CD-ROM, and in print through ham radio booksellers and regular bookstores.
- ***The QRZ DX Weekly Newsletter*** **(**`www.dxpub.com/qrz_dx_nl.html`**):** *QRZ DX* is a weekly print and e-mail newsletter with DXing news, a substantial listing of island-based (IOTA) activity, and a listing of frequencies on which sought-after stations have been heard.

The OPDX Bulletin (`www.papays.com/opdx.html`), the ARRL's *DX Bulletin,* and *Propagation Bulletin* (`www.arrl.org/bulletins`) are free weekly e-mail bulletins containing a compendium of current and anticipated DX station activities.

Contesting

Contesting activity occurs every weekend, so if you want to join in, you need to have a good event calendar at your fingertips. As you learn to enjoy the sport of contesting, you'll want to know more about operating techniques and find out how the regulars make those big scores. Here are some resources that can speed your journey from Little Pistol to Big Gun.

- ***Contest Update*** **(**www.arrl.org/the-arrl-contest-update**):** This biweekly e-mail newsletter provides news, contest results, technical topics, and product-release information of interest to high-performance hams. It's free to ARRL members.
- **Contesting.com (**www.contesting.com**):** Contesting.com is a web portal that specializes in contesting and hosts several e-mail reflectors.
- ***National Contest Journal*** **(**www.ncjweb.com**):** A bimonthly magazine published by the ARRL, *National Contest Journal* includes contest results and articles about contests and interviews. It also sponsors several popular contests.

Try these sites if you're looking for contest calendars with listings of upcoming events: www.arrl.org/contests, www.hornucopia.com/contestcal, and www.sk3bg.se/contest.

Satellites

To make contact through any of the satellites whizzing around up there, you need timely information about their status and orbitals. Getting started isn't nearly as difficult as you may think, especially if you can rely on the good how-to references listed here. The books are available through AMSAT or the ARRL.

- ***ARRL Satellite Handbook:*** This handbook, by Steve Ford (WB8IMY), offers detailed information about all aspects of satellite operation, from basic explanations to a detailed guide to satellite orbits and subsystems.
- ***Getting Started with Amateur Satellites:*** Updated in 2012, this book, by G. Gould Smith (WA4SXM), shows the beginner how to get started on satellites. It explains the necessary astronomical terminology, shows you how to locate satellites, and describes how to set up a satellite-capable station with inexpensive equipment.
- **Heavens-Above (**www.heavens-above.com**):** This site is handy for amateur use, providing information about what kinds of things you can see (or hear) in the sky. The site has a special section on ham radio satellites, as well as astronomical information on planets, comets, and other interesting phenomena.
- **Radio Amateur Satellite Corporation (**http://ww2.amsat.org**):** This site provides the latest information on satellite status and links to useful information about satellite operating.

Mobile operation

Do you prefer to do your hamming on the go? Here are some resources to answer all your questions about mobile operation.

- **A Website for Mobile Radio Operators** (`www.k0bg.com`)**:** The title of this site, by Alan Applegate (KØBG), pretty well sums up this online encyclopedia. Applegate shares his extensive background on all parts of mobile operating, from batteries to antennas.
- ***Amateur Radio Mobile Handbook:*** Published by the Radio Society of Great Britain (RSGB), this book, by Peter Dodd (G3LDO), provides an introduction to mobile operating, including how to set up a mobile station and operating guidelines.
- ***Amateur Radio on the Move:*** Written by a collection of experienced authors, this book from the ARRL covers operating not only from your car, but also from RVs, boats, and motorcycles, and even "pedestrian mobile" from a backpack.

Technical References

Ham radio involves many technologies, and to perform at optimum level, you need to know as much as you can about all of them. The following sections detail some of the best resources for many ham-related technologies.

Radio technology and electronics

Ham radio being the techie hobby that it is, you'll have more success if you can access technical references on a regular basis. Start with *The ARRL Handbook* and keep the web links in your browser bookmark file. If you decide to go beyond the basic electronics knowledge required to get your license, many good texts are available.

- ***The ARRL Handbook:*** Published by the ARRL and known simply as "the handbook" by hams and professionals alike, a copy is a must for every shack. It's an encyclopedia of the technical aspects of amateur radio and includes numerous construction projects.
- ***Batteries in a Portable World:*** Portable and emergency communication depend on batteries. This very readable book, by Isidor Buchmann, tells you all about battery technology, as well as how to choose the right battery, and how to charge and discharge batteries.

- ***Circuitbuilding Do-It-Yourself For Dummies:*** If you're just getting started and need some basic instruction about how to build those neat circuits shown in magazines and on websites, I wrote this book (available from John Wiley & Sons, Inc.) just for you.
- ***Hands-On Radio Experiments*, Volumes 1 and 2:** I wrote these books as well. Each volume contains 60 experiments or tutorials on fundamental technologies that make ham radio go. Topics range from basic electrical laws to common circuits, antennas, and feed lines.
- **K1TTT Technical Reference (**`www.k1ttt.net/technote/techref.html`**):** The K1TTT site offers numerous articles and links to information about the technical aspects of station building and design, with heavy emphasis on high-performance operating and antennas.
- ***RF Components and Circuits:*** This book, like all other books by Joe Carr (K4IPV), provides good introductory-level design and construction techniques. This particular book is a good introduction to RF circuit design and components.
- **Technical Information Service (**`www.arrl.org/technical-information-service`**):** This site is an online reference covering many useful topics and containing links to *QST* articles for ARRL members.
- ***33 Simple Weekend Projects for the Ham, the Student, and the Experimenter*:** This book, by Dave Ingram (K4TWJ), is full of good starter projects and useful gadgets to have around the ham shack and workbench. It's out of print, but used copies are commonly available at hamfests.
- ***Understanding Basic Electronics*, 2nd Edition:** This book, by Walter Banzhaf (WB1ANE), provides bite-size lessons that start at the ground floor of electronics. It discusses simple circuits and components such as the transistor and op amp.

Antennas

Hams are more likely to build antennas than any other piece of ham radio equipment. These references contain many classic and innovative designs.

- ***Antenna Zoning for the Radio Amateur:*** This guide, by Fred Hopengarten (K1VR), covers in detail the process of working with local zoning agencies to obtain permits for amateur antennas and towers.
- ***ARRL Antenna Book*, 22nd Edition:** Published by the ARRL, this book is another ham radio classic, with a thorough makeover in 2012. It covers

everything from basic antenna and transmission line theory to propagation and advanced antenna design. Useful construction projects round out every chapter, and handy software tools are provided on a CD-ROM.

- **ARRL *Classics* and *Compendium* series:** The ARRL publishes the *Antenna Classics* and *Antenna Compendium* book series, which consist primarily of construction project articles. Subjects include everything from simple wire antennas to complex arrays, microwave dishes, and transmission lines.
- ***Backyard Antennas:*** *Backyard Antennas*, by Peter Dodd (G3LDO), is a great book on getting good performance from compact antennas that fit in limited space — perfect for the suburban or urban ham.
- **Mark's RF Circuit Building Blocks Page (**`www.qsl.net/w/wa1ion`**):** Mark Connelley (WA1ION) specializes in receiving antennas and electronics for ham and SWL antennas. His website has many design articles and references to other useful sites, as well as a list of electronics design links.

VHF/UHF/microwave

Operation above 50 MHz is one of the fastest-growing areas of ham radio, with more and more excellent equipment and components becoming available every month. The following references help you assemble a working station and use it effectively.

- **DX World VHF/UHF Propagation (**`http://dxworld.com/index2.html`**):** This site is a mini portal aimed at VHF/UHF operators. It features a real-time, worldwide propagation chat screen and lots of links to useful operating and propagation resources, with a dedicated page for each of the popular bands.
- **Meteor Scatter (**`www.pingjockey.net`**):** This website, hosted by Chris Cox (NØUK), is a place for meteor-scatter enthusiasts to share ideas and propagation tips and to discuss the latest technology.
- ***6 Metre Handbook:*** Known as the "magic band" for its many propagation modes and surprise openings, 50 MHz is a favorite of many operators. This book, written by one of the band's biggest fans, Don Field (G3XTT), shows you how to join the fun.
- ***VHF/UHF Handbook,* 2nd Edition, and *The International Microwave Handbook:*** Published by the RSGB and edited by Andy Barter (G8ATD), these books take you beyond repeaters to the world of operating and wireless technology on the bands above 30 MHz.

Propagation

Propagation, which is truly fascinating, affects every ham. The more you know about it, the more interesting it becomes and the more success you'll have on the air. These references were selected to introduce the basics of radio propagation to new hams.

- **ARRL Propagation Bulletin** (`www.arrl.org/w1aw-bulletins-archive-propagation`)**:** This weekly bulletin about HF propagation is free to ARRL members.
- **HFRadio.org** (`http://prop.hfradio.org`)**:** HFRadio.org is a comprehensive site focused on HF propagation. It includes forecasts, discussions on ongoing events, and numerous charts and graphs showing historical propagation behavior.
- ***The New Shortwave Propagation Handbook:*** This handbook, by George Jacobs, Theodore J. Cohen, and Robert B. Rose, covers HF propagation from introductory levels to advanced topics. George Jacobs (W3ASK) wrote *CQ* magazine's "Propagation" column for many years.
- **SpaceWeather** (`www.spaceweather.com`)**:** This site offers news and information about the Sun and the ionosphere, and includes links to real-time satellite photos and other information. You can subscribe to the site's e-mail alerts and data updates.

Try these sites if you're looking for listings of propagation test beacons: `www.keele.ac.uk/depts/por/50.htm`, `www.ncdxf.org/beacons.html`, and the Ten-Ten International beacon list accessible via the Resources menu at `www.ten-ten.org`.

Magazines

The following magazines are the best of the general-interest ham radio publications. In addition to the ones listed in this section, every major organization likely has a membership magazine.

- ***CQ*** (`www.cq-amateur-radio.com`)**:** Focusing on general-interest stories, product reviews, and columns, *CQ* sponsors several major HF and VHF contests every year.
- ***CQ VHF*** (`www.cq-vhf.com`)**:** This magazine is tailored to all types of operation and propagation above 50 MHz, including the growing interest in using ham radio in scientific experiments, direction-finding, and orienteering.

- ***QEX*** (`www.arrl.org/qex`)**:** *QEX* (short for Calling All Experimenters) provides articles on state-of-the-art equipment and specialty articles of interest to technically advanced hams.
- ***QST* and *Digital QST*** (`www.arrl.org/qst`)**:** The ARRL's membership magazines have a great variety of technical, construction, and operating articles.
- ***WorldRadio Online*** (`www.worldradiomagazine.com`) **and *Popular Communications*** (`www.popular-communications.com`)**:** These monthly online magazines specialize in columns and short general-interest articles. *WorldRadio* deals with ham radio, and *PopComm* covers the whole family of general-interest wireless communications.

Vendors

The best way to become acquainted with the many ham radio vendors is to buy a copy of *CQ* or *CQ VHF* and scan the ads. The ARRL's archive of product reviews is another good source of equipment manufacturers. If you're getting ready to buy your first radio, the ARRL offers a free guide at `www.arrl.org/buying-your-first-radio` to help you sort out the many options.

The ham radio web portals eHam.net and QRZ.com both feature online equipment swap-and-shop pages. To find ham radio vendors on eBay, log on to `www.ebay.com` and then browse through the Computers & Electronics and Radios: CB, Ham, and Shortwave categories. You can find other useful gear in the Software and Gadgets & Other Electronics categories. More test equipment is listed for sale in the Business & Industrial and Test Equipment categories.

Index

• A •

• D •

• E •

• G •

• H •

• M •

• N •

• O •

• Q •

• R •

• S •

• T •

• U •

• V •

• W •

About the Author

H. Ward Silver, ham radio call sign NØAX, has been a ham since 1972, when he earned his Novice license and became WNØGQP. His experiences in ham radio led him to a 20-year career as an electrical engineer, designing microprocessor-based products and medical devices. In 2000, he began a second career as a teacher and writer, and he received the 2003 Bill Orr Technical Writing Award. In 2008, he was recognized as the Dayton Hamvention's Ham of the Year.

Silver is lead editor of the two primary amateur radio technical references, both published by the American Radio Relay League: *The ARRL Handbook* and *The ARRL Antenna Book*. He is the author of all three ARRL licensing study guides, and he writes the popular *QST* magazine columns "Hands-On Radio" and "Contest Corral." His popular e-mail newsletter, The ARRL Contest Update, reaches more than 25,000 readers twice a month. He has written two other Wiley titles as well: *Two-Way Radios and Scanners* and *Circuitbuilding Do-It-Yourself.* His most recent books are the ham radio detective mystery *Ray Tracy: Zone of Iniquity* (published on `www.lulu.com`) and the tutorial *Antenna Modeling for Beginners* (`www.arrl.org/shop`).

On the air, he enjoys contacting faraway (DX) stations, competing in radiosport competitions, building antennas, and participating in his local club and ARES emergency communications team. He is a founder of the World Radiosport Team Championship (`www.wrtc2014.org`) and president of The YASME Foundation (`www.yasme.org`). Outside ham radio, he plays the mandolin; dabbles in digital photography; and enjoys biking, camping, canoeing, and kayaking. Occasionally, he finds time to sleep.

Dedication

Thank you, Ellen, for all your encouragement, tolerance, and understanding. What I manage to achieve is impossible without your wholehearted support and kindness. When is our next road trip?

And to my feline companions, Dander and Imbroglio, whose constant supervision and observation were key to my completing this second edition, how about we get out the laser pointer and feathery-thing-on-a-stick?

Author's Acknowledgments

This book would be considerably poorer without the numerous individuals and organizations that contributed their photos and drawings. In addition, they've made their knowledge freely available on websites and blogs and videos in the best traditions of ham radio over the past century.

It's particularly important to acknowledge the generosity of the American Radio Relay League (ARRL; `www.arrl.org`), both in making material available to me for use in the book and in maintaining an enormous website of amateur radio know-how, significant portions of which are freely available to members and nonmembers alike. The loyal support of its members is an important reason why the organization has been available to facilitate, represent, and defend amateur radio for a century.

Publisher's Acknowledgments

Acquisitions Editor: Amy Fandrei

Project and Copy Editor: Kathy Simpson

Technical Editor: Kirk Kleinschmidt

Editorial Assistant: Annie Sullivan

Sr. Editorial Assistant: Cherie Case

Project Coordinator: Patrick Redmond

Cover Image: Robert Corse / iStockphoto

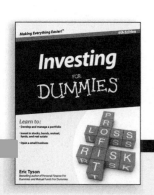

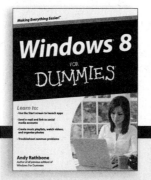

Math & Science

Algebra I For Dummies, 2nd Edition
978-0-470-55964-2

Anatomy and Physiology For Dummies, 2nd Edition
978-0-470-92326-9

Astronomy For Dummies, 3rd Edition
978-1-118-37697-3

Biology For Dummies, 2nd Edition
978-0-470-59875-7

Chemistry For Dummies, 2nd Edition
978-1-1180-0730-3

Pre-Algebra Essentials For Dummies
978-0-470-61838-7

Microsoft Office

Excel 2013 For Dummies
978-1-118-51012-4

Office 2013 All-in-One For Dummies
978-1-118-51636-2

PowerPoint 2013 For Dummies
978-1-118-50253-2

Word 2013 For Dummies
978-1-118-49123-2

Music

Blues Harmonica For Dummies
978-1-118-25269-7

Guitar For Dummies, 3rd Edition
978-1-118-11554-1

iPod & iTunes For Dummies, 10th Edition
978-1-118-50864-0

Programming

Android Application Development For Dummies, 2nd Edition
978-1-118-38710-8

iOS 6 Application Development For Dummies
978-1-118-50880-0

Java For Dummies, 5th Edition
978-0-470-37173-2

Religion & Inspiration

The Bible For Dummies
978-0-7645-5296-0

Buddhism For Dummies, 2nd Edition
978-1-118-02379-2

Catholicism For Dummies, 2nd Edition
978-1-118-07778-8

Self-Help & Relationships

Bipolar Disorder For Dummies, 2nd Edition
978-1-118-33882-7

Meditation For Dummies, 3rd Edition
978-1-118-29144-3

Seniors

Computers For Seniors For Dummies, 3rd Edition
978-1-118-11553-4

iPad For Seniors For Dummies, 5th Edition
978-1-118-49708-1

Social Security For Dummies
978-1-118-20573-0

Smartphones & Tablets

Android Phones For Dummies
978-1-118-16952-0

Kindle Fire HD For Dummies
978-1-118-42223-6

NOOK HD For Dummies, Portable Edition
978-1-118-39498-4

Surface For Dummies
978-1-118-49634-3

Test Prep

ACT For Dummies, 5th Edition
978-1-118-01259-8

ASVAB For Dummies, 3rd Edition
978-0-470-63760-9

GRE For Dummies, 7th Edition
978-0-470-88921-3

Officer Candidate Tests, For Dummies
978-0-470-59876-4

Physician's Assistant Ex
For Dummies
978-1-118-11556-5

Series 7 Exam For Dummies
978-0-470-09932-2

Windows 8

Windows 8 For Dummie
978-1-118-13461-0

Windows 8 For Dummie
Book + DVD Bundle
978-1-118-27167-4

Windows 8 All-in-One For Dummies
978-1-118-11920-4

Available in print and e-book formats.

Available wherever books are sold. For more information or to order direct: U.S. customers visit www.Dummies.com or call 1-877-762-2
U.K. customers visit www.Wileyeurope.com or call (0) 1243 843291. Canadian customers visit www.Wiley.ca or call 1-800-567-4797.
Connect with us online at www.facebook.com/fordummies or @fordummies